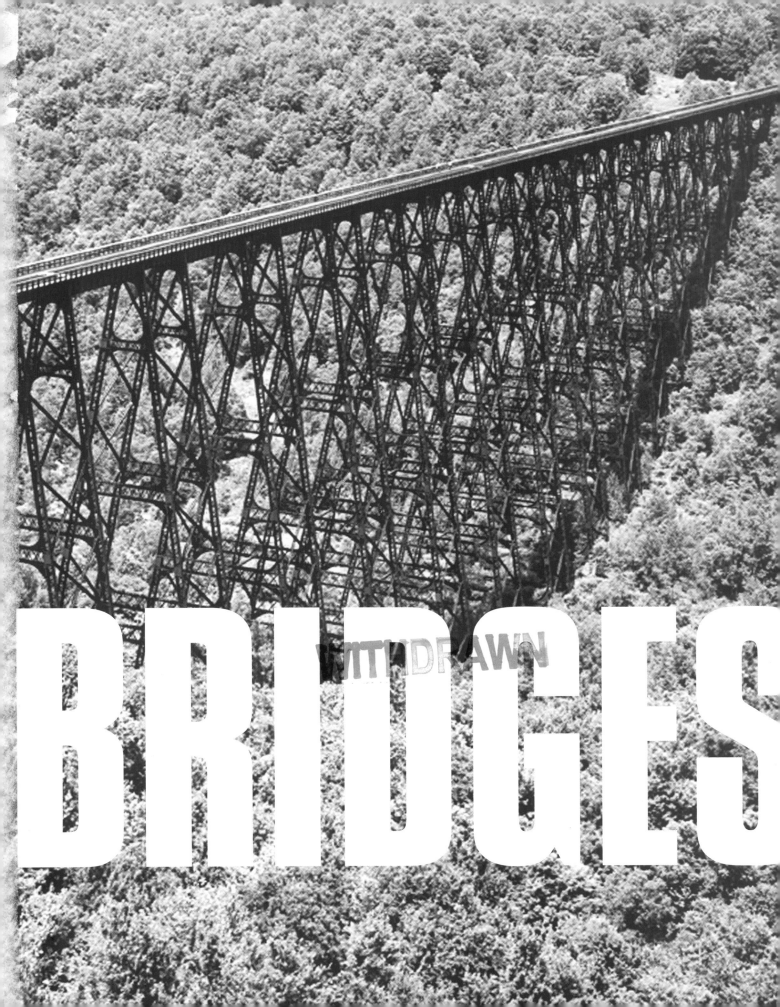

BRIDGES

NORTON / LIBRARY OF CONGRESS VISUAL SOURCEBOOKS IN ARCHITECTURE, DESIGN AND ENGINEERING

RICHARD L. CLEARY

WITH "A CALL FOR PRESERVATION" BY ERIC DELONY

W. W. Norton & Company, New York and London | Library of Congress, Washington D.C.

BRIDGES

FOR SARAH

Copyright © 2007 by Library of Congress
Copyright © 2007 Richard L. Cleary

For information about permission to reproduce selections from this book, write to
Permissions, W. W. Norton & Company, Inc., 500 Fifth Avenue, New York, NY 10110

Composition by Ken Gross
Book design by Tina Kachele
Manufacturing by Quebecor World, Peru
Production Manager: Leeann Graham
Indexing by Bob Elwood

Library of Congress Cataloging-in-Publication Data

Cleary, Richard Louis.
    Bridges / Richard L. Cleary.
        p.   cm.—(Norton/Library of Congress visual sourcebooks in architecture, design, and engineering)
    Includes bibliographical references and index.
    ISBN-13: 978-0-393-73136-1
    ISBN-10: 0-393-73136-7
        1. Bridges—United States—Design and construction—History. 2. Bridges—United States—
History. I. Title.

TG300.C54 2006
624.2—dc22   2006048203

ISBN 13: 978-0-393-73136-1
ISBN 10: 0-393-73136-7

W. W. Norton & Company, Inc.   W. W. Norton & Company Ltd.
500 Fifth Avenue   Castle House, 75/76 Wells St.
New York, N.Y. 10110   London W1T 3QT

0 9 8 7 6 5 4 3 2 1

CONTENTS

The Center for Architecture, Design and Engineering and the Publishing Office of the Library of Congress are pleased to join with W. W. Norton & Company to publish the pioneering series of the Norton/Library of Congress Visual Sourcebooks in Architecture, Design and Engineering.

Based on the unparalleled collections of the Library of Congress, this series of handsomely illustrated books draws from the collections of the nation's oldest federal cultural institution and the largest library in the world, with more than 130 million items on approximately 530 miles of bookshelves. The collections include more than 29 million books and other printed materials, 2.7 million recordings, 12 million photographs, 4.8 million maps, 5 million musical items, and 58 million manuscripts.

The subjects of architecture, design, and engineering are threaded throughout the rich fabric of this vast archive, and the books in this new series will serve not only to introduce researchers to the illustrations selected by their authors, but also to build pathways to adjacent and related materials, and even entire archives — to millions of photographs, drawings, prints, views, maps, rare publications, and written information in the general and special collections of the Library of Congress, much of it unavailable elsewhere.

Each volume serves as an entry to the collections, providing a treasury of select visual material, much of it in the public domain, for students, scholars, teachers, researchers, historians of art, architecture, design, technology, and practicing architects, engineers, and designers of all kinds.

A CD-ROM accompanying each volume contains high-quality, downloadable versions of all the illustrations. It offers a direct link to the Library's online, searchable catalogs and image files, including the hundreds of thousands of high-resolution photographs, measured drawings, and data files in the Historic American Buildings Survey, Historic American Engineering Record, and, eventually, the recently inaugurated Historic American Landscape Survey. The Library's Web site has rapidly become one of the most popular and valuable locations on the Internet, experiencing over 3.7 billion hits a year and serving audiences ranging from school children to the most advanced scholars throughout the world, with a potential usefulness that has only begun to be explored.

Among the subjects to be covered in this series are building types, building materials and details; historical periods and movements; landscape architecture and garden design; interior and ornamental design and furnishings; and industrial design. *Bridges* is an excellent exemplar of the goals and possibilities on which its series is based.

JAMES H. BILLINGTON
THE LIBRARIAN OF CONGRESS

The introduction to this book provides an overview of the construction of bridges in the United States. It is a view that is broad and inspired by the depth and quality of the resources of the Library of Congress. The balance of the book, containing 919 images, which can be found on the CD at the back of the book, is organized into five sections. Figure-number prefixes designate the section.

Captions give the essential identifying information, when known: subject, location, designer, builder, and/or engineer, creator(s) of the image, date, and the Library of Congress call number, which can be used to find the image online. Note that a link to the Library of Congress Web site may be found on the CD.

ABBREVIATIONS USED IN CAPTIONS

DETR	Detroit Publishing Company Collection
FSA	Farm Security Administration
Gen. Coll.	General Collection
HABS	Historic American Building Survey
LC	Library of Congress
P&P	Prints and Photographs Division
G&M	Geography and Map Division
RBSCD	Rare Books and Special Collections Division
HAER	Historic American Engineering Record
NYWTS	New York World Telegram and Sun Collection

BRIDGE BUILDING IN AMERICA

Bridges are such ubiquitous features of the built environment that we cross most of them barely acknowledging their presence (IN-001). Certain bridges, however, command our attention. Some stand out simply for their utility facilitating travel from here to there; others stir our imaginations by their size, setting, beauty, or historical associations. Ordinary or spellbinding, every bridge is a response to a problem—the spanning of a river or other obstacle, solved more or less elegantly. Accounts of bridges in guidebooks and Web sites tend to emphasize statistical achievements, such as dimensions or traffic capacity, but the measure of a bridge's significance as a creative solution often is revealed most clearly in details, such as the connections of structural members.

IN-001. Judith River Bridge, Moore Vic., Fergus County, Montana, 1912. Security Bridge Company (Billings, MT), builder. Jet T. Lowe, photographer, 1980. P&P,HAER, MONT,14-MOR.V,1-2.

IN-001

IN-002

This visual sourcebook surveys the history of bridge design in the United States in terms of four fundamental structural types (beam, arch, truss, and suspension) and the special functional category of movable bridges (swing, lift, and bascule—that is, a bridge with one or more deck sections hinged to tilt upward). The typological organization allows us to consider how similar structural ideas have been addressed by different designers, refined over time, and rendered in various building materials. Alongside familiar monuments of American bridge building appear many modest structures known only locally, at best, but which offer equally eloquent statements of problems solved.[1]

IN-002. Map of the Great Central Route showing Railroad Suspension Bridge, Niagara River, Niagara Falls, New York, 1855. John A. Roebling, engineer. C. E. Noble (Buffalo, NY), publisher, 1856. G&M,G3701.P3 1856.N6 RR 421.

BRIDGES AND AMERICAN TRANSPORTATION HISTORY

Innovation in bridge design since the founding of the United States has been tied closely to the transportation policies of government and industry. Sustained investment in highways and canals intended to facilitate settlement and commerce across the newly established nation began in the 1790s.[2] The Philadelphia & Lancaster Turnpike Road (begun 1792) and other initiatives at the end of the eighteenth century were followed by more ambitious projects including the National Road from Cumberland, Maryland, to Wheeling, West Virginia (built 1811–1817, extended to Vandalia, Illinois, by 1841), and the Erie Canal (begun 1817). In 1828, the Baltimore & Ohio Railroad received its charter as the first railroad in the United States and began construction of a line from Baltimore to the Ohio River. Forty-one years later, railroads spanned the continent and were the primary mode of long-distance inland shipping (IN-002).[3] The success of all these endeavors required reliable bridges.

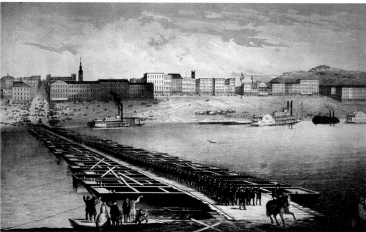

IN-003

IN-004

IN-003. "Destruction of the locomotives on the bridge over the Chickahominy." Drawing by Alfred R. Waud, 1862. P&P,LC-USZC4-5821.

Waud, one of the foremost Civil War sketch artists, recorded the destruction of a bridge in Virginia in this drawing for Harper's Weekly. The notation above the image reads, "The R. R. Bridge was burned. Trees too high."

IN-004. "The Pontoon Bridge at Cincinnati." Lithograph by Middleton, Strobridge & Company (Cincinnati, OH) after a drawing by A. E. Matthews, ca. 1862. P&P,LC-USZ62-142.

During the Civil War, roads, canals, and railroads were strategic assets, and bridges were recognized as vulnerable links (IN-003, IN-004).[4] Military engineers on both sides of the conflict became proficient at rapidly replacing destroyed structures with lightweight wooden trusses and pontoon bridges, but the work often was hazardous. An account by Colonel Wesley Brainerd of the 50th New York Volunteer Engineers describes the construction of a pontoon bridge during the Battle of Fredericksburg in 1862 (IN-005):

> We had completed the tenth abutment or length, making the bridge 200 ft. long (the Stream at this point was about 400 ft. wide). I was standing at the extreme outer end of the bridge encouraging my men, when, happening to cast my eyes to the shore beyond just as the fogg [*sic*] lifted a little, I saw, what for the moment almost chilled my blood. A long line of arms moving rapidly up and down was all I saw, for a moment later they were again obscured by the fog. But I knew too well that line of arms was *ramming cartridges* [itals. original] and that the crisis was near.[5]

In the ensuing engagement, every member of Brainerd's ten-man squad was killed or wounded (he was shot in the arm), but the completion of the bridge was merely delayed, not stopped. Throughout the war, Union engineers, led by figures such as Herman Haupt (1817–1905), possessed the organization and resources to outbuild their Confederate counterparts.

Over the course of the nineteenth century, the rapidly increasing numbers of shippers and travelers dependent on reliable transportation networks became less tolerant of delays at ferry crossings of rivers once considered beyond the capabilities of bridge builders. Designers responded by developing economical structural systems (including the suspension bridge), materials, and methods of assembly that made it possible to span longer distances and to build intermediate piers in deep river channels. Successively,

IN-005. "Building pontoon bridges at Fredericksburg Dec. 11th." Drawing by Alfred R. Waud, 1862. P&P,LC-USZC4-3604.

Waud's drawing shows members of the 50th New York Engineers Regiment building a pontoon bridge during the Battle of Fredericksburg, Virginia.

IN-005

bridges carrying road and rail traffic spanned the Susquehanna, Ohio, Mississippi, and other major rivers.

Heavier vehicular loads were another spur to innovation. From the 1840s to the 1890s, the weight of railroad locomotives increased fivefold from around twenty tons to over one hundred tons.[6] Bridge designers had to address both the greater total weight and its uneven distribution as the heavy locomotive moved across the span. With the exception of some masonry arch bridges, structures adequate for one generation of rolling stock had to be reinforced or replaced to accommodate the increased loads imposed by the next. Rapid construction was of the essence, and though traditional masonry arch bridges were structurally efficient and durable, they required greater investments of time, labor, and materials than many railroads were willing to make as they expanded their networks in a highly competitive business environment. Wood and iron, which lent themselves to the economies of standardization and prefabrication, became the most common construction materials.[7]

Wood was readily available throughout much of the United States and easily worked. Standardized designs, such as the arch-reinforced truss patented by Theodore Burr in 1817 (see 3-017), reduced the guesswork in arranging the structural members, and a host of entrepreneurs patented variations on basic truss configurations. In proportion to its weight, wood offers high resistance to compressive and tensile loads, but it can be weakened by rot, insects, fires ignited by sparks from the fireboxes of steam locomotives, and connections loosened by repeated flexing caused by traffic and wind. In the nineteenth century, the estimated life expectancy of wooden structural elements exposed to the ele-

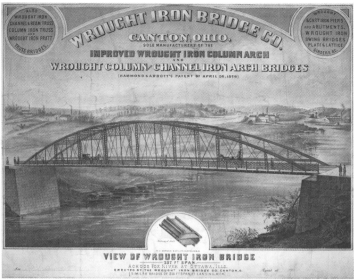

IN-006

IN-007

IN-006. Advertisement, "R. W. Smith's Patent Truss Bridge," 1866. Robert W. Smith, bridge designer, Tippecanoe, Ohio. Lithograph published by Strobridge & Company, (Cincinnati, OH) P&P,LC-USZC4-5593.

This advertisement illustrates a common story in mid-nineteenth-century bridge building. Robert W. Smith (born 1834) began his career erecting barns with innovative roof structures and running a woodworking shop and lumber yard with his brother in Tippecanoe. He expanded into bridge building in 1866 with a patented design for a wooden lattice-type truss. In 1867, he moved to Toledo and founded the Smith Bridge Company, which produced wood and, later, iron road and railroad bridges.

IN-007. Advertisement for Wrought Iron Bridge Company (Canton, OH) showing the Fox River Bridge, Ottawa, La Salle County, Illinois, ca. 1870. W. J. Morgan Company, lithographer, 1870. P&P,LC-USZC4-2354.

ments was seven or eight years, but builders had learned that it could be at least doubled by encasing trusses with a roof and siding.[8] The first covered bridge in the United States was Palmer's Permanent Bridge in Philadelphia, completed in 1805, the final link of the Philadelphia & Lancaster Turnpike Road (see 3-015). Another approach to increasing the longevity of wooden bridges was to use iron to reinforce or replace certain members and connections. The Howe truss, patented in 1840, replaced wooden tension members with wrought-iron rods and bolted connections that could be adjusted in response to shifting alignments (see 3-138). In the second half of the century, iron became the material of choice for long-span bridges, but builders have continued to use wood for trestles and other types of short spans.

The oldest surviving all-iron bridge built in the United States is the 80-foot, cast-iron arch designed by Richard Delafield in 1836 that still spans Dunlap's Creek in Brownsville, Pennsylvania (see 2-044).[9] Within a decade, builders were erecting iron trusses with compression members of cast iron and tension members of wrought iron, which was less brittle and had higher tensile strength than cast iron but was too expensive for use throughout (see 3-148). Bridge manufacturing rapidly became an important segment of the emerging iron industry, and fabricators aggressively marketed stock designs through advertising and the charm of traveling salesmen (IN-006, IN-007).[10] Mass production of steel became possible after the Civil War, and the Eads (see 2-050), Glasgow (Missouri; see 3-272), and Brooklyn (see 5-033) bridges built between 1874–83 demonstrated its efficacy in arch, truss, and suspension construction. As its quality and affordability increased in the 1880s, steel rapidly supplanted wrought iron; it remains, with reinforced concrete, one of the two most common materials for bridge construction.

IN-008

IN-008. Lincoln Highway Bridge, Mud Creek, Tama, Tama County, Iowa, 1914. Iowa State Highway Commission, designer; Paul Kingsley (Strawberry Point, IA), contractor. Joseph Elliott, photographer, 1995. P&P,HAER,IOWA,86-TAMA,1-2.

Enthused by the prospect of being on the route of the newly defined Lincoln Highway, officials in Tama embellished this otherwise undistinguished 20-foot, reinforced concrete slab bridge with elegant lamps and a railing that assured travelers that they were on the right road.

At the turn of the twentieth century, railroads dominated long-distance overland transportation. Road transport, not yet mechanized, primarily served local needs. By the 1920s, however, the automobile had gone from curiosity to necessity, and road construction had taken on the urgency formerly associated with the building of railroad lines.[11] Within a decade, projects were under way for transcontinental trunk routes, such as the Lincoln Highway (conceived 1912; IN-008), and the first limited-access divided highways, including the Merritt Parkway (begun 1934; see 1-068) in Connecticut, the Pennsylvania Turnpike (begun 1938; see IN-014), and the Arroyo Seco Parkway (begun 1938; see 2-107) in California. The heavier traffic loads in terms of volume and weight and the higher speeds of motorized transport required bridge designers not only to improve carrying capacity but also to consider driver safety issues such as sight lines and continuity in roadway alignments. Increasingly, highway bridges were understood as components of systems rather than as isolated objects.

State highway departments emerged as the primary managers of those systems, assuming responsibilities handled previously by counties.[12] Their importance increased during the Great Depression when they became funding centers for public works projects.[13] Often working in concert with the federal Bureau of Public Roads, they reviewed or provided designs for bridges and instituted statewide construction standards. This movement toward standardization coincided with ongoing consolidation of the steel industry exemplified by the formation of the United States Steel Corporation in 1901 and its subsidiary, the American Bridge Company (IN-009–IN-012). As the opportunities diminished for small-scale builders to market their proprietary designs directly to local

IN-009. Empire Bridge Company, Elmira, Chemung County, New York. Haines Photo Company (Conneaut, OH), photographer, ca. 1909. P&P, PAN US GEOG—New York no. 11 (F size).

As part of its consolidation effort, the American Bridge Company, a division of the United States Steel Corporation, operated the Empire Bridge Company from 1900 to 1914 as its center of production for bridges built in New York state.

IN-010. Ambridge, Beaver County, Pennsylvania. Arthur Rothstein, photographer, 1938. P&P, LC-USF34-026531-D.

In 1903, the American Bridge Company purchased 2,500 acres of mostly undeveloped land along the Ohio River downstream of Pittsburgh and began construction of what became the world's largest plant for steel and bridge fabrication. Two years later, the borough took the name Ambridge. By 1930, its population was over 30,000.

IN-011. Workers' housing, Ambridge, Beaver County, Pennsylvania. Arthur Rothstein, photographer, 1938. P&P, LC-USF34-026529-D.

IN-012. Millworker and family, Ambridge, Beaver County, Pennsylvania. John Vachon, photographer, 1941. P&P, LC-USF34-062172-D.

IN-010

IN-011

IN-012

governments, the variety of bridge types erected on state and county highways decreased (albeit with the benefit of improved safety records). Each state, however, exercised its own preference for certain designs based on the professional biases of its engineering staff, and this sometimes resulted in a subtle differentiation in the appearance of bridges from state to state. For example, riveted steel Warren pony and Parker through trusses were defining features of Wisconsin's highway system from the 1910s to the 1930s (see 3-233), while in Louisiana the state highway department employed the K-truss to a degree not found elsewhere (see 3-323).[14]

IN-013

IN-014

IN-013. Construction photo, Monroe Street Bridge, Spokane River, Spokane, Washington, 1911. J. C. Ralston, P. F. Kennedy, Morton MacCartney, J. F. Greene, engineers; Cutter & Malmgren (Spokane, WA), builder. W. O. Reed, photographer, August 3, 1911. P&P,LC-USZ62-121882.

The photo shows the 281-foot concrete deck arch supported by wooden centering. A cable suspended above the bridge facilitated the movement of building materials.

IN-014. Pennsylvania Turnpike Overpass, Pennsylvania, ca. 1940. Arthur Rothstein, photographer, July 1942. P&P,LC-USW3-005733-D.

The rapid expansion of the nation's highways occurred at the same time reinforced concrete gained widespread acceptance as a primary structural material, and many highway departments adopted it alongside steel. Concrete, a mixture of sand, water, cement, and aggregate, can be thought of as two materials. When simply poured into formwork or cast into blocks, it behaves like stone, strong in compression, relatively weak under tension. When steel is embedded within it, however, concrete gains tensile strength and can be used for many of the same applications as steel.

The principles of reinforced concrete construction were developed in the mid-nineteenth century in Europe and introduced to the United States after the Civil War. The first reinforced concrete bridge in the country is the Alvord Lake Bridge in San Francisco, designed in 1889 by Ernest Ransome (see 2-095). Over the next two decades, Ransome and other engineers, such as Fritz von Emperger and Daniel Luten, patented a variety of reinforcing systems and licensed them to construction companies (see 2-138, 2-146). Engineers across the country employed concrete for highway and railroad bridges, but innovation and competition were particularly intense in the Midwest and on the West Coast. During the first half of the twentieth century, engineers most commonly designed concrete spans as arches (IN-013), but by the 1920s, improved manufacturing and assembly techniques allowed the production of longer beams, cast in place or prefabricated, and concrete girder bridges gradually became the familiar sights on American highways so ubiquitous today (IN-014).

IN-015. Bridge designs, Andrea Palladio, The Architecture of A. Palladio; in Four Books, ed. Giacomo Leoni (London, 1742), vol. 1, book 3, plates III, IV. RBSCD, NA 2517 .P3 fol.

By the mid-eighteenth century, Palladio's Quattro libri d'architettura (1570) was available in a variety of editions ranging from expensive folios, such as this example owned by Thomas Jefferson, to pocket-size volumes. The section on the design of wooden bridges was the first extended treatment of the subject published during the Renaissance, and it served as a reference for builders through the first decades of the nineteenth century. Palladio illustrated his text with examples of his own work, including a bridge he built on the Cismone River (top) and designs for various configurations of trusses informed by his study of ancient Roman sources, his experience with roof trusses, and prevailing construction practices in the Veneto.

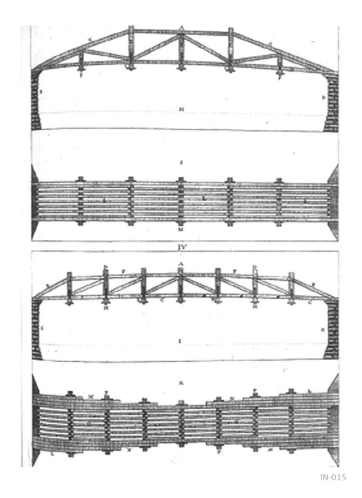

IN-015

BRIDGE DESIGN AND ENGINEERS

Transportation history identifies key forces that drive innovation in bridge design. Study of the transformation of engineering from a craft to a professional discipline helps to explain how innovation occurred. In the first decades of the nineteenth century, bridge design was based on empirical knowledge passed along by craftsmen as rules of thumb, and innovation was a matter of extrapolation. For example, Theodore Burr (1771–1822; see 3-017) extended his knowledge of heavy-timber framing acquired as a millwright to bridges and derived his influential combination of the arch and the truss from his direct knowledge of the two structural systems.[15] High-style architects possessing private libraries had access to a wider variety of examples than craftsmen through illustrations in books, such as Andrea Palladio's *Quattro libri d'architettura* (1570, available in English-language editions since 1738; IN-015), but similarly relied on experience when predicting the behavior of their designs.

By the 1820s, Stephen Long (1784–1864) and a few other American engineers had begun to apply quantitatively derived theories formulated in Europe regarding the behavior of beams and other aspects of mechanics to the design of trusses. Squire Whipple (1804–1888) codified this knowledge in 1847 with his book *A Work on Bridge*

Building, which redefined bridge engineering in America as a discipline based on scientific principles.[16] Its appearance was timely, because the insistent demands for innovation imposed by the railroads could not be realized without a critical framework within which new ideas could be developed and evaluated.

As its scientific basis gained acceptance, so did engineering's reputation as a learned profession. Engineers founded professional societies, including the American Society of Civil Engineers (1852). Colleges and universities, informed by European polytechnical schools, introduced curricula for training future engineers. The U.S. Military Academy began offering formal courses in engineering in 1813, and Rensselaer Polytechnic Institute created the nation's first engineering degree in 1835. In 1862, Congress passed the Morrill Act, which assisted states in establishing colleges (the "land grant" universities) dedicated to teaching the applied sciences of agriculture and the mechanical arts.

Although formal academic training became increasingly important during the nineteenth and early twentieth centuries, many American engineers would continue to enter the profession through a combination of informal study and work experience until professional registration laws standardized admission criteria (Wyoming passed the first in 1907). James Eads (1820–1887), the designer of the pioneering iron and steel arch bridge (1867–1874) that bears his name in St. Louis (see 2-050), acquired his expertise through independent reading and his experiences with underwater salvage and the construction of ironclad warships. Octave Chanute (1832–1910), best known today for the contributions he made to aeronautical science late in his career, worked his way up through the ranks of railroad engineering. He began at the age of sixteen as a chainman on a surveying crew; fifteen years later, he was named chief engineer of the Chicago & Alton Railroad with responsibility for the design of terminals, stockyards, and other structures, as well as bridges (see 4-004).

Chanute's position as the chief engineer of a railroad represents just one of the many profesional niches engineers have occupied among the conceptual, technical, and economic processes involved in bridge design and construction. The extremes of this range are marked on one end by the craftsman who builds the structures he designs, and on the other by the consulting engineer retained to review a specific aspect of a project under development. Other roles include entrepreneurs promoting patented designs, such as Joseph Strauss (1870–1938), chief engineer of the Golden Gate Bridge (see 5-095), who had made his reputation in the first decades of the twentieth century devising mechanisms for bascule bridges (see 4-081); and manufacturers, such as the Roebling family, who produced wire rope for over a century (IN-016–IN-018).

IN-016. Wire Mill No. 2, Kinkora Works, John A. Roebling's Sons Company, Roebling, Burlington County, New Jersey, after 1904. Charles G. Roebling, designer. Joseph Elliott, photographer, 1997. P&P,HAER,NJ,3-ROEBL,1B-17.

IN-017. Site plan in 1930, Kinkora Works, John A. Roebling's Sons Company, Roebling, Burlington County, New Jersey. Charles G. Roebling, designer. Amy Wynne, Thomas Behrens, delineators, 1997. P&P,HAER,NJ,3-ROEBL,1-,sheet no. 3.

John A. Roebling located his wire rope company in Trenton in 1848. In 1904, his sons purchased a tract of farmland on the Delaware River about 10 miles south of the city and established a new plant that allowed the firm to manufacture its own steel and wire, which would be made into cables for suspension bridges and other uses at the Trenton works. Charles G. Roebling's (1849–1918) efficient layout of production facilities included a village that housed 40 percent of the workforce. The Roebling family retained control of the plant until 1953.

IN-018. Building styles, Kinkora Works, John A. Roebling's Sons Company, Roebling, Burlington County, New Jersey. Charles G. Roebling, designer. Amy Wynn, Thomas Behrens, delineators, 1997. P&P,HAER,NJ,3-ROEBL,1-,sheet no. 14.

Charles G. Roebling supervised the design of all the plant's buildings from foundries to housing.

IN-016

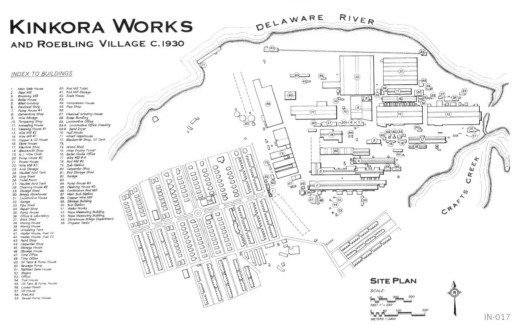

KINKORA WORKS
AND ROEBLING VILLAGE C.1930

INDEX TO BUILDINGS

IN-017

ROEBLING BUILDING STYLES

Charles Roebling was responsible for the layout and design of both the manufactory and the worker's village. As a result, there are strong stylistic similarities between the industrial architecture of the plant and the civic architecture of the town. These drawings are quick surveys intending to illustrate the similarities.

GENERAL STORE

2ND AVENUE ROW HOUSES

BLACKSMITH SHOP

DUPLEX HOUSE

ROEBLING HOUSE

ROEBLING INN

MACHINE SHOP

POWER HOUSE

ROEBLING AUDITORIUM

ROEBLING PUBLIC SCHOOL

WIRE MILL NUMBER 2

ROEBLING BOARDING HOUSE

IN-018

These niches represent more than job descriptions, for they often constitute subcultures within the broader engineering profession. Each favors certain approaches to problem solving, has its own institutional orthodoxies, and involves a degree of collaboration rarely suggested by the design credits, which typically are assigned to the chief engineer alone.[17] In engineering and in architecture, the design of long-span bridges, tall office buildings, or other complex structures requires a supporting cast of players who apply their talents to realizing a vision orchestrated by the principals. Unique, innovative projects typically require larger staffs than those based on standardized designs. Gustav Lindenthal, for example, employed a staff of ninety-five for the design of the Hell Gate Bridge in New York (1904–1917; see 2-073), while at about the same time the Phoenix Bridge Company's staff of thirty-five adapted stock plans for projects across the country (IN-019–IN-021).[18] Today, computerized drafting and computation allow fewer people to accomplish tasks that once required many hands, but the work remains a team effort, and, as in the past, the young engineers and architects moving through the ranks of public agencies and private firms master the nuances of the prevailing subcultures and form professional networks that disseminate particular viewpoints.

The significant role such subcultures play in the history of bridge engineering is evident in the fortunes of innovative technologies and design principles. A noteworthy aspect of the history of reinforced concrete as a primary structural material for bridges in the United States is the degree to which innovation occurred in the Midwest and West. One center of activity in the first decades of the twentieth century was Iowa. James Barney Marsh (1856–1936), whose distinctive "rainbow arch" reinforced concrete bridges were once common sights in midwestern towns (see 2-153), based his Marsh Engineering Company in Des Moines and employed engineering graduates of Iowa State University in Ames, who were well prepared to design with the still relatively new material.[19] The university also provided engineers for the Iowa State Highway Department, which was an early adapter of reinforced concrete construction. An alumnus of all three organizations was Conde B. McCullough (1887–1946), who made his career in Oregon designing elegant reinforced concrete bridges in the 1920s and 1930s as chief bridge engineer for the state's rapidly expanding highway system (see 2-164). McCullough staffed his department with graduates of Iowa State and Oregon State University (then Oregon Agricultural College), where he had taught before joining the highway department.[20] We can thus identify an engineering subculture transmitted from Iowa to Oregon. Although the theories shared by this group were widely available to anyone through publications, their acceptance in practice relied greatly on the direct interaction of individuals brought together by the circumstances of school or employment.

IN-019. Foundry and workers' housing, Phoenix Iron and Steel Company, Phoenixville, Chester County, Pennsylvania, 1882 (foundry). Amory Coffin, architect. Jet T. Lowe, photographer, 1999. P&P,HAER,PA,15-PHOEN,4A-1.

A subsidiary of the Phoenix Iron and Steel Company, Phoenix Bridge was one of the most prolific bridge manufacturers at the end of the nineteenth century. It integrated production from design through fabrication and erection, and it marketed its products internationally via catalogs and an aggressive sales force. The firm lost market share in the twentieth century and closed in 1962. Some facilities at the Phoenixville works remained in use until 1984.

IN-020. Foundry interior, Phoenix Iron and Steel Company, Phoenixville, Chester County, Pennsylvania, 1882 (foundry). Amory Coffin, architect. Jet T. Lowe, photographer, 1999. P&P,HAER,PA,15-PHOEN,4A-11.

IN-021. Girder Shop No. 6, interior. Phoenix Iron and Steel Company, Phoenixville, Chester County, Pennsylvania. Jet T. Lowe, photographer, 1999. P&P,HAER,PA,15-PHOEN,4D-5.

IN-022

IN-022. Collapse, Tacoma Narrows Bridge, Puget Sound, Tacoma, Pierce County, Washington, 1938–1940. Leon Moisseiff, engineer, superstructure; Clark H. Eldridge, engineer, piers; Pacific Bridge Company, Bethlehem Steel Company, builders. James Bashford, photographer, November 7, 1940. P&P,LC-USZ62-46682.

In engineering, as in other professions, commonly accepted theories and practices may prove to be flawed, sometimes with disastrous results. A classic instance of this phenomenon was the collapse of "Galloping Gertie," the Tacoma Narrows Bridge torn apart by wind-induced oscillations on November 7, 1940, just four months after it had opened (IN-022).[21] Only one life was lost—a dog named Tubby—but the financial cost was high, and the failure of a structure designed by leading engineers and hailed as state-of-the-art attracted close scrutiny. The underlying cause was neither a calculation error by the design team nor a mistake by the contractors or suppliers. Indeed, the bridge's design and construction had followed prevailing standards of practice. The problem resided in a set of theoretical assumptions known as deflection theory, which had been embraced by the subculture of long-span suspension bridge designers as the key to a new generation of longer spans with ribbon-thin decks. Published by the Austrian engineer Josef Melan (1853–1941) in 1888, it had been introduced to the United States at the turn of the twentieth century by Leon Moisseiff (1872–1943) in the design of the Manhattan Bridge (see 5-066), in New York City. The enthusiastic acceptance of the theory overshadowed critical investigations of its limitations informed by analysis of the behavior of older bridges, and the lesson of its limitations was learned the hard way. The Tacoma bridge was rebuilt (see 5-083) with deeper, stiffer trusses, and the decks of other bridges designed according to deflection theory, such as the Bronx-Whitestone Bridge in New York City (see 5-104), were reinforced similarly.

IN-023

The attraction of the thin, elegant lines of the suspension bridges designed according to deflection theory was in part aesthetic, an aspect of bridge design that has been at times a matter of controversy both within the engineering profession and among engineers, architects, and artists (IN-023). The professional distinction between architecture and engineering crystallized in the nineteenth century. In its most extreme characterization, the division ceded beauty to architects and functional efficiency to engineers; but from the nineteenth century to the present, there have been architects who place a premium on expediency and engineers who view their work as an art.

What makes a bridge beautiful? Many engineers and architects alike would agree that the answer begins with a structurally efficient design that clearly expresses its underlying principles.[22] However, for a given design problem, there often is more than one solution meeting these criteria. Are all beautiful? On this question, opinions have varied greatly. Some have maintained that good bridge design is framed above all by practicality, safety, and economy; others argue that a beautiful design is one that responds to additional factors, such as the character of the site or the social circumstances of the commission.

Around the turn of the twentieth century, bridge projects in urban centers began including architects on the design teams.[23] This added expense was justified in the spirit of the City Beautiful movement, which maintained that well-designed bridges, water pumping stations, and other utilitarian structures contributed to a sense of civic identity along with more traditional public monuments such as courthouses and museums. In this spirit, a number of prominent architects, such as Henry Hornbostel (see 2-072) at

IN-024

the beginning of the century and Cass Gilbert (IN-024) and Paul Cret (IN-025) in the 1920s and 1930s, took a special interest in lending their expertise to bridge design. The scope of work allotted to architects varied from the design of ornamental details to consulting on overall character. A frequent assignment was the articulation of abutments, piers, and towers, the components most closely related to architectural structures.

The marriage of the two professions, neither of which was predisposed to play a supporting role for the other, was not without conflict, and in some instances the contributions of architects seemed dispensable. The masonry cladding Cass Gilbert designed for the steel towers of the George Washington Bridge (1923–1931) in New York City was not executed for financial reasons, but the omission was viewed positively by some who felt that covering the steel structure was dishonest and out of step with the spirit of modernity.[24] Indeed, by the 1930s, the question of what expertise architects could bring to bridge design had become considerably more complicated than it had been earlier in the century when their task was understood as beautifying structures according to the widely accepted principles and forms of classical art. In contrast, modernist doctrines of

IN-024. George Washington Bridge, Hudson River, New York, New York, 1927–1931. Othmar H. Ammann, chief engineer; Allston Dana, design engineer; Cass Gilbert, architect. Margaret Bourke-White, photographer, ca. 1940. P&P,LC-USZ62-83229.

IN-025. Market Street Bridge, Susquehanna River, Harrisburg, Pennsylvania, 1926–1928. Ralph Modjeski and Frank N. Masters, engineers; Paul Cret, architect; James McGraw Company, contractor. Joseph Elliott, photographer, 1997. P&P,HAER, PA,22-HARBU,27-4.

good design variously called for forms that would be uniquely expressive of a new age, informed by the rational thinking of science and engineering, and based on concepts that eliminated distinctions between form and structure. While turn-of-the-century architects were summoned to bring style to engineers, the latter were now seen as capable guides for architects who had lost their way in historicism. Despite the changing polemics, the role of architects in bridge projects has not been reduced to the design of tollbooths, and architects and engineers alike have continued to view the aesthetics of bridges as a subject relevant to both professions.[25]

BRIDGES AND SOCIETY

Like buildings and other prominent structures, bridges can have a commanding presence in the landscape that invites us to take notice of our surroundings. Some bridges have been planned for just such a purpose. London Bridge, for example, the nineteenth-century successor of the medieval bridge falling down in the nursery rhyme, was trans-

IN-026

IN-026. London Bridge, Thames River, London, England, 1824–1831. Jolliffe and Banks, designers. Engraving published by J. McCormick (London), 1831. P&P,LC-DIGppmsca-02629.

IN-027. London Bridge, Lake Havasu City, Mohave County, Arizona, built in London 1824–1831; relocated 1968–1971. Jolliffe and Banks, designers; Robert Beresford, reconstruction. Jet T. Lowe, photographer, 1998. P&P,HAER,ARIZ,8-LANAC,1-1.

The bridge was reconstructed with a concrete core faced with original stones.

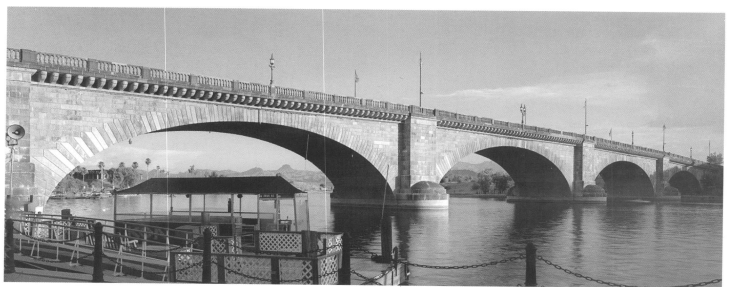

IN-027

ported to Arizona in 1968 and reconstructed as a magnet for real estate development (IN-026, IN-027). Advertisers have long exploited the power of bridges as attention-getting devices to promote the sale of such diverse products as sewing machines, railroad tickets, and automobiles (IN-028–IN-030). Artists, too, have responded to their scale and character, sometimes critically (IN-023). The writer Hart Crane was so moved by the Brooklyn Bridge as an expression of American spirit that he made it the subject of an epic poem, *The Bridge* (1930).

In addition to their properties as objects, bridges have an inherent relational quality not always evident in architecture. Bridges link previously separate places and alter our perception of them. The appearance of a river can be changed profoundly by the construction of a bridge that laces together the parallel ribbons of its banks and channel

IN-028. Advertisement for the Empire Sewing Machine Company showing the Brooklyn Bridge, New York. Lithograph published by Henry Seibert & Brothers, ca. 1870. P&P, LC-USZC4-6492.

IN-029. Advertisement for Oldsmobiles showing the Benjamin Franklin Bridge, Philadelphia, Pennsylvania. Published in Look magazine, February 18, 1958. P&P, LC-USZC4-2495.

The advertising team for Oldsmobile apparently had none of Joseph Pennell's reservations (IN-025) regarding the beauty of the bridge.

IN-030. Advertisement for the New York Central Lines showing the Alfred H. Smith Memorial Bridge, Castelton-on-Hudson, Rensselaer County, New York. Lithograph by Latham Lithography & Printing Company, 1924. P&P, LC-USZC4-9720.

IN-028

IN-029

IN-030

(IN-031). Moreover, the link is not limited to the span itself. Whether it supports a road, railroad track, or pipeline, a bridge usually is part of a network that continues beyond its immediate setting. Remote as its location may be, it implies the existence of other places down the road.

Designers have addressed the issue of the bridge as an integral element of a landscape in several ways. One is exemplified by the National Park Service's "rustic style" of the 1930s, which employed finish materials and, to a lesser extent, structural type to echo features of the site (see 2-134). Other designers, such as Conde B. McCullough, have

IN-031. Panoramic view of bridges spanning the Des Moines River, Des Moines, Polk County, Iowa. Hebard-Showers Company, photographer, 1919. P&P,LC-USZ62-109334.

IN-032. Golden Gate Bridge, San Francisco Bay, San Francisco, California, 1933–1937. Joseph B. Strauss, chief engineer; O. H. Ammann, Charles Derleth Jr., Charles Ellis, Leon S. Moisseiff, consulting engineers; Irving F. and Gertrude C. Morrow, consulting architects. Bethlehem Steel, superstructure; John Roebling's Sons Company, cables; Pacific Bridge Company, substructure. Jet T. Lowe, photographer, 1984. P&P,HAER, CAL,38-SANFRA,140-47.

taken less literal approaches, seeking instead to achieve a dialogue between complementary forms of bridge and landscape.

"What Nature rent asunder long ago, man has joined today," proclaimed Joseph Strauss, chief engineer of the Golden Gate Bridge at its opening in 1937 (IN-032).[26] The presence of a bridge as a link can inspire our imaginations to dwell on themes of connection that go beyond its physical properties. Coupled with the significant investments of money and effort, the symbolic associations of a bridge make its completion an opportunity for public celebrations (IN-033, IN-034) and commemoration in song (IN-035; see also 5-029, 5-038). On such occasions, civic leaders have mined the symbolic aspect of bridges bringing people together. In such a spirit, a fragment from Shake-

IN-033

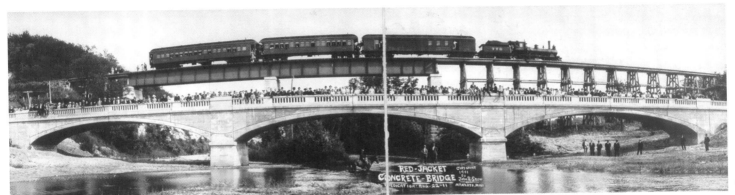

IN-034

speare's *Hamlet* was paraphrased at the inauguration of the Wheeling Suspension Bridge in 1849: "The friends thou hast and their adoption tried, Grapple them to thy soul with hooks of steel."[27] In the hands of songwriters, the theme has been pursued in more intimate terms: "So give to me the pathway sweet, Above the waters blue," rhapsodized the author of "Strolling on the Brooklyn Bridge," in 1883; "It's on the Brooklyn Bridge! We meet, I and my darling true."[28] A vision of artistic, if not intentionally romantic, meetings informed the design of the architect Michael Graves in 1977 for a cultural center that would span the Red River linking the twin cities of Fargo, North Dakota, and Moorhead, Minnesota (IN-036). Another expression of a bridge fostering public interaction is the Bridge of Flowers in Shelburne Falls, Massachusetts, a former trolley bridge transformed into a garden in 1929 (IN-037).

Bridges also serve as sites of public memory. Most commonly, this ambition is limited to the selection of a name and a dedicatory plaque, but on occasion it has been a

IN-035

IN-035. "The Wheeling Bridge Polka," composed by John Fickeisen, published by Firth, Pond and Company, New York, 1853. Music Division, M1.A12I vol. 53 Case Class original bound volumes.

Above the picture of the bridge is the name of its designer, Charles Ellet.

factor shaping design. Bridges planned according to the precepts of the City Beautiful movement projected civic values as well as good taste and conveyed their messages with sculpture and inscriptions. The towers at the ends of the Michigan Avenue Bridge in Chicago (1920) display allegorical sculpture representing stages of the city's settlement (see 4-062). A more militant theme appears in the pylons, sculpture, and inscriptions on the Soldiers' and Sailors' Memorial Bridge in Harrisburg, Pennsylvania (1930), which honors the state's veterans and forms part of a monumental approach to the capitol (see 2-120). In both examples, the character of the iconographic programs is reinforced by the cladding of the structural members (steel in Chicago, reinforced concrete in Harrisburg) with high-quality stone masonry, the traditional material of choice for monumental civic architecture. This vision of monumentality was mostly abandoned after World War II owing to different budgetary priorities, design considerations related to higher traffic speeds and volumes, and a new aesthetic climate that favored unadorned structures. Today, designers charged with creating memorable bridges turn to expressive structural types such as arches and cable-stayed spans (see 5-118).

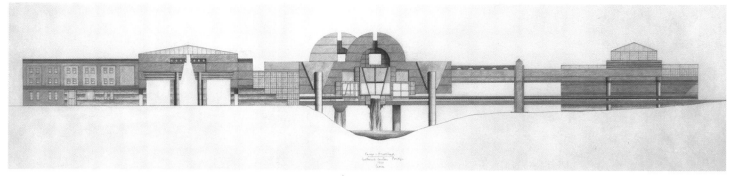

IN-036. Fargo-Moorhead Cultural Center Bridge project, Red River, Fargo, Cass County, North Dakota, and Moorhead, Clay County, Minnesota, 1978. Michael Graves, architect. Drawing. P&P,LC-USZC4-10448.

IN-037. Bridge of Flowers, Deerfield River, Shelburne Falls, Franklin County, Massachusetts, built 1908, converted 1929. Albertype Company, photographer. P&P, Vert file: Bridges—MA; SSF.

In 1929, local residents transformed this former trolley bridge into a public garden filled with flower beds, and it has become the town's leading attraction.

IN-037

Upon completion, a bridge may seem destined to stand for the ages, but weather and use can degrade its structural integrity and increased traffic loads can render it obsolete. Historically, most bridges have been regarded as disposable commodities to be replaced as circumstances demand. Yet every bridge embodies both a way of thinking realized through a particular technology and certain cultural associations as an object in a particular place. Herein arises a challenge in assessing significance so considered decisions can be made whether to rehabilitate a bridge, demolish it, or remove it to a protected location. To this end, statewide surveys of historic bridges have been undertaken throughout the country, and more detailed documentation of select structures has been made by programs such as the Historic American Engineering Record (HAER), whose photographs, drawings, and reports deposited in the Library of Congress provide an extraordinary resource for historic preservation issues and a comprehensive history of the technical, social, and economic dimensions of American bridge building.

NOTE

The illustrations survey the variety of bridge types found across the United States. They show key features of bridge construction and topics discussed in this introduction, including the character of bridges as features in the landscape (grand or humble) and examples of the roles bridges play in American culture.

The names of bridges used for the captions are those given in the Library of Congress collections, and the locations are as cited therein. When new locations have been identified, they are given as well. County names are included for all locations except those within the boundaries of a major city. Single dates refer to the date of completion. If a bridge is known to have been demolished, it is so indicated. Readers should be aware that HAER documentation of many bridges occurred as part of mitigation agreements in advance of planned demolition, but the reports often do not record the final disposition of a structure.

In the credits for design and construction, the term "engineer" is used for members of that profession (trained by apprenticeship or formal study) responsible for the design of the bridge; the term "builder" refers to an individual (who may be an engineer) or firm responsible for its construction. The addresses (city and state) of builders are included when available so readers may trace the roles of local and national contracting firms and manufacturers on given projects. In the twentieth century, as today, large companies such as the American Bridge Company had multiple offices and manufacturing plants; these locations, rather than the corporate headquarters, often appear in project documentation.

This book is a portal. Its illustrations and topics are meant to inspire readers to pursue the study of individual bridges and the history of bridge building in America. A first step is to examine the HAER reports linked to the photographs available through the Library of Congress Web site. Many of the reports are definitive, but historians and engineers continue to refine the analysis and interpretation of historic bridges. Examples of this scholarship are offered in the Bibliography.

ACKNOWLEDGMENTS

The historians, engineers, and architects studying historic bridges are remarkable for their generosity. I thank Eric DeLony for introducing me to this community and for his unwavering support. For their critical readings of chapter drafts, I thank Mark Brown, Sarah Cleary, James L. Cooper, Bob Frame, David Guise, Robert L. Hadlow, Emory L.

Kemp, and Steve Olson. C. Ford Peatross and the excellent staff I encountered in every department facilitated my research at the Library of Congress. I also wish to acknowledge the assistance of graduate research assistant Kristen Brown and the many specialists who patiently responded to my queries, including Joseph D. Conwill, Dario Gasparini, Robert Jackson, Cheryl Maze and the Figg Engineering Group, Robert McCullough, Ann L. Miller, Antony Opperman, Barbara Pruitt and HNTB Engineering and Architecture, Stephen Roper, and David Simmons. Nancy N. Green and Andrea Costella capably guided the book's production. The Page Southerland Page Fellowship in Architecture supported research expenses.

NOTES

1. For a concise survey of American bridges organized chronologically, see Eric DeLony, *Landmark American Bridges* (New York: American Society of Civil Engineers, 1993). For a history organized by building material, see David Plowden, *Bridges: The Spans of North America* (New York: W. W. Norton, 2002).

2. George Rogers Taylor, *The Transportation Revolution, 1815–1860*. Economic History of the United States, vol. 4 (New York: Holt, Rinehart and Winston, 1951), Chaps. 2–3.

3. Ibid., Chap. 5.

4. See George Edgar Turner, *Victory Rode the Rails: The Strategic Place of the Railroads in the Civil War* (1953; Lincoln: University of Nebraska Press, 1992).

5. Wesley Brainerd, *Bridge Building in Wartime: Colonel Wesley Brainerd's Memoir of the 50th New York Volunteer Engineers*, ed. Ed Malled (Knoxville: University of Tennessee Press, 1997), 113.

6. John H. White, *American Locomotives: An Engineering History, 1830–1880* (Baltimore: Johns Hopkins Press, 1968), 21–22. Edwin P. Alexander, *American Locomotives: A Pictorial Record of Steam Power, 1900–1950* (New York: Bonanza Books, 1950), 22.

7. George Danko, "The Evolution of the Simple Truss Bridge 1790 to 1850: From Empiricism to Scientific Construction" (Ph.D. dissertation, University of Pennsylvania, 1979), 112–16.

8. Octave Chanute, "Statement and Estimates Accompanying Plans for a Bridge across the Misouri [sic] River at St. Charles," manuscript signed and dated August 14, 1865 (Octave Chanute Collection, Department of Manuscripts, Library of Congress).

9. Frances C. Robb, "Cast Aside: The First Cast-Iron Bridge in the United States," *IA*, Journal of the Society for Industrial Archeology, 19, no. 2 (1993): 48–62.

10. For an account of the marketing strategy of one of the country's leading bridge manufacturing companies during the second half of the nineteenth century, see David A. Simmons, "Bridge Building on a National Scale: The King Iron Bridge and Manufacturing Company," *IA*, Journal of the Society for Industrial Archeology, 15, no. 2 (1989): 23–39. For a survey of the promotional literature published by bridge manufacturers, see Victor C. Darnell, "The Other Literature of Bridge Building," *IA*, Journal of the Society for Industrial Archeology, 15, no. 2 (1989): 40–56.

11. Bruce E. Seely, *Building the American Highway System: Engineers as Policy Makers* (Philadelphia: Temple University Press, 1987), 72–74.

12. The histories of the state highway departments of Iowa and Oregon illustrate this development. See Robert W. Hadlow, *Elegant Arches, Soaring Spans: C. B. McCullough, Oregon's Master Bridge Builder* (Corvallis: Oregon State University Press, 2001), 20–29, 42–43, 87–89.

13. Seely, *Building the American Highway System*, 88–93.

14. John N. Vogel, "CTH D Bridge, Spanning Eau Galle River, Eau Galle vicinity, Dunn County, WI," Historic American Engineering Record Report HAER No. WI-100 (1997), 6–7, Library of Congress; Paul Hawke, "Krotz Springs Bridge, Spanning Atchafalaya River, Krotz Springs, St. Landry Parish, LA," Historic American Engineering Record Report HAER No. LA-7 (1983), 5, Library of Congress.

15. Danko, "The Evolution of the Simple Truss Bridge," 60ff.

16. Ibid., 158, 219.

17. Louis Bucciarelli, "An Ethnographic Perspective on Engineering Design," *Design Studies* 5, no. 3 (1984): 185–90. Eda Kranakis has examined bridge design in terms of engineering subcultures in *Constructing a Bridge: An Exploration of Engineering Culture, Design, and Research in Nineteenth-Century France and America* (Cambridge, MA: MIT Press, 1997) and "Fixing the Blame: Organizational Culture and the Quebec Bridge Collapse," *Technology and Culture* 45, no. 3 (2004): 487–518.

18. For Lindenthal's staff, see Henry Petroski, *Engineers of Dreams: Great Bridge Builders and the Spanning of America* (New York: Vintage Books, 1996), 189; for the size of the Phoenix Bridge Company's engineering staff, see Kranakis, "Fixing the Blame," 500.

19. Hadlow, *Elegant Arches, Soaring Spans*, 18–19.

20. Ibid., 44, 46.

21. Petroski, *Engineers of Dreams*, 294–308.

22. For a discussion of bridge aesthetics in a current design handbook, see M. S. Troitsky, *Planning and Design of Bridges* (New York: John Wiley & Sons, 1994), Chap. 10. See also Fritz Leonhardt, *Brücken/Bridges: Aesthetics and Design* (Cambridge, MA: MIT Press, 1984).

23. See Jonathan Farnham, "Staging the Tragedy of Time: Paul Cret and the Delaware River Bridge," *Journal of the Society of Architectural Historians* 57, no. 3 (September 1998): 258–79.

24. Petroski, *Engineers of Dreams*, 264–65.

25. See Building Arts Forum/New York, *Bridging the Gap: Rethinking the Relationship of Architect and Engineer*, proceedings of the Building Arts Forum/New York symposium, Guggenheim Museum, April 1989, ed. Deborah Gans (New York: Van Nostrand Reinhold, 1991).

26. Quoted in Allen Brown, *Golden Gate: Biography of a Bridge* (New York: Doubleday & Co., 1965), 126.

27. Based on *Hamlet*, 1. iii. 61; reported in *The Daily Wheeling Gazette*, November 17, 1849, p. 2; article reprinted http://wheeling.weirton.lib.wv.us/landmark/bridges/susp/opening.htm.

28. "Strolling on the Brooklyn Bridge," lyrics by George Cooper, music by J. P. Skelly (New York: Richard A. Saalfield, 1883). Library of Congress, Music Division, M2.3.U6A44.

A CALL FOR
PRESERVATION

Though the American Society of Civil Engineers (ASCE) was responsible for helping create the Historic American Engineering Record (HAER) in the late 1960s, few engineers at the time practiced or embraced the preservation of their own heritage. Historic preservation then was a new concept, with only a few having any clue of its meaning—much less ramifications for the future. Over the past forty years, the movement has spread. Today, it generally is recognized that aspects of our cultural heritage are worth preserving, including bridges.

Following the collapse of the Silver Bridge into the Ohio River in Point Pleasant, West Virginia, in 1967—the worst bridge disaster of the twentieth century—congressional action took place to rid the nation's highways of functionally deficient and structurally obsolete bridges. Unfortunately, this well-intentioned legislative action inadvertently threatened to eliminate all evidence of the engineers' art from the American landscape. Thus, it was with a sense of urgency that HAER's historic bridge program began in 1975. Over a thousand bridges have been recorded since then.

Though there is evidence of some progress in saving historic bridges, it is not enough. Recent statistics suggest that over half the historic bridges of the United States were destroyed in the last two decades of the twentieth century—two decades in which preservation awareness of structures of all types was at its most sophisticated level.[1] Despite this sophistication, as a nation we have yet to resolve how to save representative examples of our historic bridges. Until there is a national policy with specific legislative and funding incentives to do so, bridges remain a heritage at risk.

Indeed, bridges not only illustrate economic development and engineering prowess, but as this book reveals, they are indelibly linked to the cultural landscape of this vast country. Most people recognize the engineering icons like the Brooklyn Bridge, the Golden Gate, or the steel and wrought-iron arches of James Buchanan Eads's magnificent span across the mighty Mississippi. Some people resonate to nostalgic covered bridges and stone arches, while others embrace the Erector-set metal truss or solid concrete arches. While America has more covered bridges than any other country—about 750—the true bridge heritage at risk comprises the iron trusses and concrete arches fabricated during the late nineteenth and early twentieth centuries by bridge companies and sold through catalogs. Hundreds of patents were granted, and no other country experimented with the truss form or concrete arch as did the United States. Americans depended on these structures to tie their communities together and link them to larger cities. In fact, many of these bridges remain, some still carrying traffic. Many have passed the century mark, and some are wearing out.

Congress mandated a national historic bridge program in 1987. Now that program has to be backed up with funding sources for preserving historic bridges and leadership by highway departments at local, state, and federal levels and engineering firms that possess the knowledge to care for these structures. Some engineering firms are beginning to carve out expertise in the rehabilitation of historic bridges as part of their everyday practice, and a number of state departments of transportation are recognizing that the preservation of historic bridges is part of their responsibility for comprehensive highway planning.[2] They are finding that it is possible to bring additional excellence to America's highways by combining new construction with preservation of the very best handed down to us from the past, which we in turn can pass on to the future.

A new challenge in this effort is the assessment and selective preservation of the thousands of bridges built in the decades following World War II as part of the nation's Interstate Highway System (Dwight D. Eisenhower National System of Interstate and Defense Highways), which celebrated its 50th anniversary on June 29, 2006. Historians have only recently begun to study the steel beams, cantilevers, concrete slabs, and reinforced and prestressed concrete girders that highway departments developed for overpasses, short-, and mid-length spans.[3] The majority of bridges and other structures (including interchanges, tunnels, and rest areas) located within the Interstate Highway System's 46,700 mile right-of-way have been exempted from eligibility for designation in the National Register of Historic Places, but certain features of national or exceptional significance will remain eligible for consideration as historic properties.

Saving historic bridges of fine materials, humanly scaled proportions, notable craftsmanship, and varied textures enhances the quality of life and maintains familiar surroundings. It also makes economic sense as Americans and foreign visitors alike discover the fascinating matrix of highways and scenic byways interlaced with the farm-to-market roads that knit this country together.

Future generations will judge us by how well we succeeded in saving examples of our historic built environment, including historic bridges, as we enter the twenty-first century. No one is advocating that every example be saved, but certainly there is no argument against the preservation of selected spans. Some can be continued in vehicular service. For others, the solution may be relocation to lesser-used roads, trailways, or bikeways.

In places where it is perceived that historic architectural and cultural resources are lacking, attitudes supporting good design may also be absent. Such values are especially needed in America where we tend to throw away the past, pursue the quick profit, and in the name of progress, irrevocably alter the landscape. In such an environment, it is critical to remember that historic bridges provide a link with the past as well as deeper insights and hope for the future.

Eric DeLony, Chief (Retired)
Historic American Engineering Record
National Park Service

NOTES

1. Eric DeLony and Terry H. Klein, *Historic Bridges: A Heritage at Risk, A Report on a Workshop on the Preservation and Management of Historic Bridges, Washington, DC, December 3–4, 2003.* SRI Preservation Conference Series 1, June 2004: p. 18 (http://www.srifoundation.org/ index.html). Probably half the bridges illustrated in this book have been destroyed, though without analysis, exact numbers are not available.

2. In addition to citizen involvement and leadership at state and local levels, we need a national historic bridge program. This only can come from the federal level—that is, through the Federal Highway Administration.

3. Reinforced and prestressed concrete and steel slabs, beams, and girders have become the most common bridge types, numbering in the thousands and superceding the once ubiquitous single-intersection Pratt and Warren through and pony trusses. Parsons Brinkerhoff, Engineering & Industrial Heritage, PC, *A Context for Common Historic Bridge Types.* National Cooperative Highway Research Program, Transportation Research Council, Washington, D.C., 2005. (http://www4.trb.org/trb/crp.nsf/reference/boilerplate/attachments/$file/25-25(15)_FR.pdf).

BEAM BRIDGES

THE SIMPLEST STRUCTURE for a bridge is a beam, such as a plank or log crossing

a stream. The principal beams of a bridge are known as girders, and the spans

they form are categorized according to three types. A simple span is a beam

supported at its ends. A continuous span has intermediate supports, like a

rock supporting a log mid-stream, and a cantilevered span extends the beam

beyond its support, like a diving board.

1-001. Members of the Primitive Baptist Church assembling for a baptism, Morehead, Rowan County, Kentucky. Marion Post Wolcott, photographer, 1940. P&P,LC-USF33-031002-M3.

Planks supported by rocks at their ends form a series of simple spans. The weight of persons crossing the bridge imposes a vertical load on the beams, causing them to deflect. A beam's resistance to deflection is a function of its material properties, unsupported length, and depth.

1-001

1-002

1-003

1-004

1-005

1-002. Log footbridge, Jackson Vic., Breathitt County, Kentucky. Marion Post Wolcott, photographer, 1940. P&P,LC-USF33-031133-M1.

A pair of logs serve as girders forming a simple beam. Smaller, split logs support a plank deck.

1-003. Prescott Bridge, Lamprey River, Raymond, Rockingham County, New Hampshire, 1917. United Construction Company (Albany, NY), designer and builder. Bruce Alexander, photographer, 1988–89. P&P,HAER,NH-8,RAYM,1-4.

A pair of 40-foot, 27-inch-deep steel I-beam girders form this simple span.

1-004. South San Gabriel River Bridge, Georgetown, Williamson County, Texas, 1939. Texas Highway Department, designer; Dean Word (San Antonio, TX), builder. Bruce A. Harris, photographer, 2001. P&P,HAER,TX,246-GEOTO,2-3.

Individual beam segments cantilevered beyond the piers are riveted to a center beam, forming a continuous span.

1-005. Detail of riveted connection joining cantilevered beams and center beam, South San Gabriel River Bridge, Georgetown, Williamson County, Texas, 1939. Texas Highway Department, designer; Dean Word (San Antonio, TX), builder. Bruce A. Harris, photographer, 2001. P&P,HAER,TX,246-GEOTO,2-6.

STONE CRIB
CONSTRUCTION

1-006. Bailey Island Bridge, Casco Bay, Bailey Island, Cumberland County, Maine, 1928. Llewellyn N. Edwards, engineer. Jet T. Lowe, photographer, 1984. P&P,HAER, ME,3-BAILI,1-11.

Granite has extraordinary strength under compressive loads, which makes it and other types of stone well suited for building walls and piers; but it is proportionally much less resistant to tensile forces, limiting its use for beams to short spans, as seen here. Llewellyn Edwards designed this unusual bridge as a series of cribs with the locally quarried granite blocks laid to allow the tide to ebb and flow.

1-007. Detail of granite cribs, Bailey Island Bridge, Casco Bay, Bailey Island, Cumberland County, Maine, 1928. Llewellyn N. Edwards, engineer. Jet T. Lowe, photographer, 1984. P&P,HAER,ME,3-BAILI,1-13.

1-008

1-009

1-010

WOODEN CONSTRUCTION

1-008. Sewell's Bridge, York River, York, York County, Maine, 1761 (demolished 1934). Samuel Sewell, builder. DETR, photographer, ca. 1908. P&P,LC-D4-70310.

For crossings over relatively shallow water, wooden piles may be driven into the earth at intervals commensurate with the bearing capacity of the beams. Pile and beam construction was introduced to America from Europe during the colonial era, and variants of this structural system continue to be built today.

1-009. Lithograph, North Eastern Railroad Trestle, Charleston, South Carolina, ca. 1854. William Keenan, lithographer, ca. 1854. P&P,LC-USZ62-14863.

1-010. Annisquam Bridge, Lobster Cove, Gloucester, Essex County, Massachusetts, 1861 (rebuilt 1946, demolished ca. 1987). Joseph B. Burnham, original builder. Jeffrey C. Howry, photographer, 1986. P&P,HAER, MASS,5-GLO,4-1.

1-011. Grapevine Bridge, Chickahominy River, Virginia, May 27–28, 1862. Built by 5th New Hampshire Infantry. David B. Woodbury, photographer, 1862. P&P,LC-B8171-7383.

1-012. Ogden-Lucin Cutoff Trestle, Southern Pacific Railroad, Great Salt Lake, Brigham City Vic., Box Elder County, Utah, 1902–1904. William Hood, chief engineer; Southern Pacific Railroad, designer. Jack E. Boucher, photographer, 1971. P&P,HAER,UTAH,2-BRICI,3-34.

Pilings organized as frames (bents) distinguish trestle bridges from simple pile and beam construction. This trestle is nearly 12 miles long and crosses a portion of the Great Salt Lake.

1-013. Cutaway view, Ogden-Lucin Cutoff Trestle, Southern Pacific Railroad, Great Salt Lake, Brigham City Vic., Box Elder County, Utah, 1902–1904. William Hood, chief engineer; Southern Pacific Railroad, designer. Robert J. McNair, delineator, 1971. P&P,HAER, UTAH,2-BRICI,3-,sheet no. 5.

An unusual feature of this trestle is the ballasted roadbed consisting of a bed of crushed rock (ballast) that absorbs vibrations and provides a quieter ride for passenger trains.

1-011

1-012

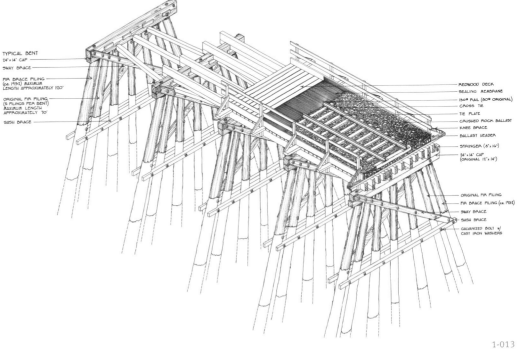

1-013

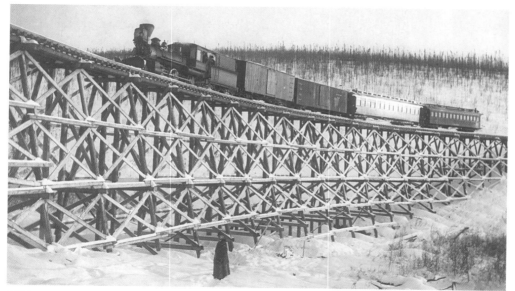

1-014

1-015

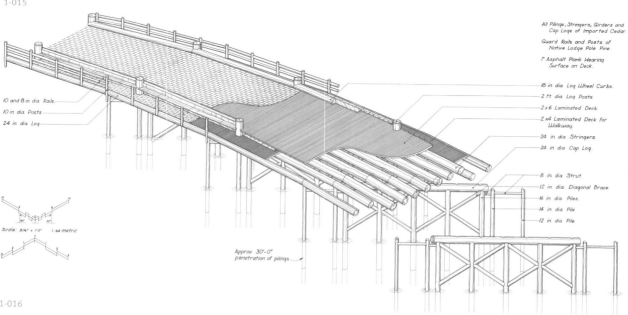

1-016

1-014. Fox Gulch Bridge, Tanana Valley Railroad Company, Fox Vic., Fairbanks North County, Alaska. Albert J. Johnson, photographer, 1916. P&P,LC-USZ62-88755.

1-015. Fishing Bridge, Yellowstone River, Yellowstone National Park, Teton County, Wyoming, 1937. W. G. Carnes, National Park Service, architect; G. M. Williams, Bureau of Public Roads, engineer; Strong & Grant (Springville, UT), builder. Jet T. Lowe, photographer, 1983. P&P,HAER,WYO,15-YELNAP, 3-10.

1-016. Cutaway view, Fishing Bridge, Yellowstone River, Yellowstone National Park, Teton County, Wyoming, 1937. W. G. Carnes, National Park Service, architect; G. M. Williams, Bureau of Public Roads, engineer; Strong & Grant (Springville, UT), builder. Julie E. Pearson, Laura E. Salarano, delineators, 1989. P&P,HAER,WYO,15-YELNAP, 3-,sheet no. 3.

Due to the alignment of the road, the bridge spans the river at an oblique angle to the current. To minimize the exposure of the bents, or trestles, to the force of the current and floating objects, the builders skewed them with respect to the axis of the bridge.

1-017. Outlet Creek Bridge, Lake Sullivan, Colville National Forest, Metaline Falls Vic., Pend Oreille County, Washington, 1935. U.S. Forest Service, designer; Civilian Conservation Corps, builder. Harvey S. Rice, photographer, 1993. P&P,HAER,WASH,26-METFA.V, 1-1.

1-018. 'Auwaiakeakua Bridge, 'Auwa-iakeakua Gulch, Waikoloa, Hawaii County, Hawaii, 1940. William R. Bartels, Territorial Highways Department, designer; Otto Medeiros, builder. David Franzen, photographer, 1997. P&P,HAER,HI,1-WAIK,1-5.

From the mid-nineteenth century to the 1890s, when high-quality steel became afford-able and widely available, girders typically were made of wrought iron.

1-019

1-020

1-019. Bow Bridge, Central Park, New York, New York, 1859. Calvert Vaux, designer. Jet T. Lowe, photographer, 1984. P&P,HAER, NY,31-NEYO,153A-5.

Decorative iron bridges were introduced as features in European gardens in the late eighteenth century. They were technological novelties that could be erected quickly and allowed designers great flexibility for stylistic expression. Calvert Vaux (1824–1895) and his assistant, Jacob Wrey Mould (1825–1886), adopted this practice for Central Park. Five of their bridges remain today (1-020; see also 2-047–2-049).

1-020. View of Pinebank Arch during renovation, Central Park, New York, New York, 1861. Calvert Vaux, designer; J. B. & W. W. Cornell Foundry (New York, NY), fabricator. Jet T. Lowe, photographer, 1984. P&P,HAER,NY,31-NEYO,153B-2.

Like Bow Bridge, Pinebank Arch structurally is a wrought-iron girder, shaped for aesthetic reasons as a shallow arch.

1-021. El Capitan Bridge, Merced River, Yosemite National Park, Mariposa County, California, 1933. Bureau of Public Roads, designer; Sullivan & Sullivan, contractor. Brian C. Grogan, photographer, 1991. P&P,HAER,CAL,22-YOSEM,22-1.

1-022. Construction details, El Capitan Bridge, Merced River, Yosemite National Park, Mariposa County, California, 1933. Bureau of Public Roads, designer; Sullivan & Sullivan, contractor. Marie-Claude Le Sauteur, delineator, 1991. P&P,HAER,CAL,22-YOSEM,22-,sheet no. 2.

Logs on the exterior elevations mask the steel girder structure and express the "rustic style" favored by the National Park Service in the 1930s.

1-021

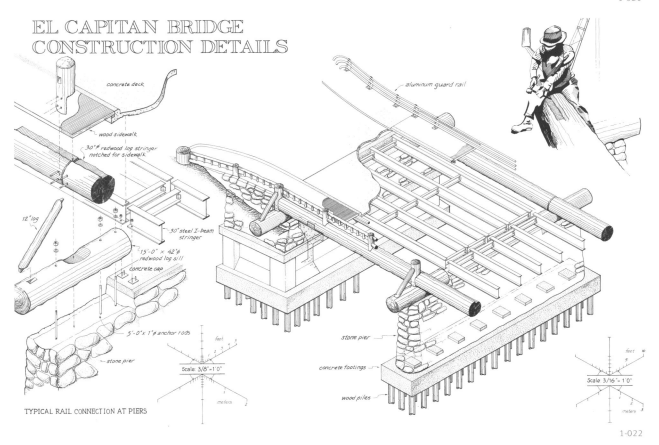

EL CAPITAN BRIDGE
CONSTRUCTION DETAILS

concrete deck

wood sidewalk

30"∅ redwood log stringer
notched for sidewalk

12" log

30" steel I-beam
stringer

15'-0" x 42"∅
redwood log sill

concrete cap

5'-0" x 1"∅ anchor rods

stone pier

aluminum guard rail

stone pier

concrete footings

wood piles

Scale: 3/8"=1'0"

Scale: 3/16"=1'0"

TYPICAL RAIL CONNECTION AT PIERS

1-022

1-023

1-024

1-025

1-023. Grand Avenue Viaduct, Sioux City, Iowa, 1936. Ash, Howard, Needles & Tammen (Kansas City, MO), engineers; C. F. Lytle Company (Sioux City, IA), contractor. Bruce A. Harms, photographer, 1996. P&P,HAER, IOWA,97-SIOCI,4-11.

This cantilevered, continuous-span, steel girder viaduct carries highway traffic over stockyards and railyards for nearly 4,000 feet.

1-024. Gibbon River Bridge No. 1, Yellowstone National Park, Park County, Wyoming, 1938. W. G. Carnes, National Park Service, and Bureau of Public Roads, designers; Stone & Grant (Springville, UT), builder. Jet T. Lowe, photographer, 1989. P&P,HAER,WYO,15-YELNAP,5-1.

1-025. Detail, Gibbon River Bridge No. 1, Yellowstone National Park, Park County, Wyoming, 1938. W. G. Carnes, National Park Service, and Bureau of Public Roads, designers; Stone & Grant (Springville, UT), builder. Jet T. Lowe, photographer, 1989. P&P,HAER, WYO,15-YELNAP,5-5.

The outer girders of this continuous-span structure are encased with concrete.

1-026

1-027

1-028

1-026. Tygart Valley Homesteads Bridge, Randolph County, West Virginia, 1938–1939. John Vachon, photographer, June 1939. P&P,LC-USF34-060070-D.

The Tygart Valley Homesteads was a community of 165 housing units and light industry planned by the federal Division of Subsistence Homesteads as part of New Deal programs intended to improve economic conditions in rural areas during the Great Depression. The photograph contrasts the modern steel girder bridge with the old, "submarine" ford beside it, which was impassable in times of high water.

1-027. Construction photo, Tygart Valley Homsteads Bridge, Randolph County, West Virginia, 1938–1939. Marion Post Wolcott, photographer, September 1938. P&P,LC-USF33-030127-M1.

1-028. Construction photo showing riveted connections, Tygart Valley Homsteads Bridge, Randolph County, West Virginia, 1938–1939. Marion Post Wolcott, photographer, September 1938. P&P,LC-USF33-030127-M4.

1-029. Cantilever Bridge, New Jersey Turnpike, New Jersey, 1950–1952. New Jersey Turnpike Authority, designer. Gottscho-Schleisner, photographer, 1952. P&P,LC-G613-62216.

During the rapid expansion of the nation's highway system following World War II, cantilevered, continuous-span, steel, deck-girder bridges and their reinforced concrete counterparts were adopted by state highway departments across the country for overpasses and other short and mid-length spans.

1-029

PLATE GIRDER CONSTRUCTION

Plate girders are composed of wrought-iron or steel segments riveted or welded together. Their fabrication and transportation can be more economical than that of large, rolled beams.

1-030

1-031

1-030. West Falls Bridge, Philadelphia & Reading Railroad, Schuylkill River, Philadelphia, Pennsylvania, 1890. H. K. Nichols, chief engineer, Philadelphia & Reading Railroad; Pencoyd Bridge & Construction Company, fabricator; Nolan & Brothers Company, contractor. Joseph Elliott, photographer, 1999. P&P,HAER,PA,51-PHILA,726-1.

The bridge includes an 80-foot stone arch and eight plate girder spans of 60 to 92 feet.

1-031. Pecos River High Bridge (Pecos Viaduct), Southern Pacific Railroad, Langtry Vic., Val Verde County, Texas, 1892 (demolished 1949). Julius Kruttschnitt, Southern Pacific Railroad, engineer; Phoenix Bridge Company (Phoenixville, PA), fabricator. Benton Scooper, photographer, ca. 1922. P&P,LC-USZ62-93083.

The roadbed of this 2,180-foot steel trestle bridge stood 321 feet above the river, making it the tallest railroad bridge in North America when it was built.

1-032. Cincinnati, Jackson & Mackinaw Railroad Bridge, Thom's Run, Farmersville Vic., Montgomery County, Ohio, ca. 1895–1900. Kimberly Starbuck, photographer, 1994. P&P,HAER,OHIO,57-FARMV.V,2-5.

In the late nineteenth century, some engineers favored "fishbelly" girders, shaped to be deepest at the middle where the bending stresses are highest, as an efficient solution for accommodating the heavy loads of railroad traffic. The central span of this bridge is 72 feet.

1-032

1-033

1-034

1-035

1-033. Aerial view, Randolph-Wells Street Station, Union Elevated Railroad, Chicago, Illinois, 1897. J. A. L. Waddell, engineer. Jack E. Boucher, photographer, 1971. P&P,HAER,ILL,16-CHIG,108F-1.

This station on the Chicago Loop railway designed by one of the nation's leading engineers at the turn of the twentieth century is elevated above the street by a system of riveted steel columns, plate girders, and trusses.

1-034. Street-level view, Randolph-Wells Street Station, Union Elevated Railroad, Chicago, Illinois, 1897. J. A. L. Waddell, engineer. Jack E. Boucher, photographer. P&P,HAER,ILL,16-CHIG,108F-6.

1-035. Kinzua Viaduct, New York, Lake Erie & Western Railroad, Kinzua Creek Valley, Mount Jewett Vic., McKean County, Pennsylvania, 1882, rebuilt 1900 (collapsed 2003). 1882 viaduct: Octave Chanute, Oliver W. Barnes, NYLA&W engineers; Adolphus Bonzano, engineer, Phoenixville Bridge Works Company; Phoenix Bridge Company (Phoenixville, PA), builder. 1900 viaduct: C. W. Bucholz, Octave Chanute, Mason Strong, engineers; Elmira Bridge Company, builder. Jack E. Boucher, photographer, 1971. P&P,HAER, PA,42-MOJEW.V,1-17.

The single-track, iron, lattice deck girder viaduct of 1882 was 2,053 feet long and 302 feet high. It was replaced in 1900 by the similar but stronger steel structure with plate deck girders, shown here, which collapsed in a windstorm in 2003.

1-036. Kate Shelley Memorial High Bridge (Chicago & Northwestern Railroad viaduct), Des Moines River, Boone Vic., Boone County, Iowa, 1901. E. C. Carter, W. H. Finley, C&NW, engineers; George S. Morison, consulting engineer; American Bridge Company, fabricator. Joseph Elliott, photographer, 1995. P&P,HAER,IOWA,8-BOONE.V,1-15.

A series of plate-girder spans and a subdivided Pratt deck truss carry a pair of tracks 2,685 feet across the Des Moines River Valley.

1-036

1-037

1-037. Kate Shelley Memorial High Bridge (Chicago & Northwestern Railroad viaduct), Des Moines River, Boone Vic., Boone County, Iowa, 1901. E. C. Carter, W. H. Finley, C&NW, engineers; George S. Morison, consulting engineer; American Bridge Company, fabricator. William Henry Jackson, photographer, ca. 1901. P&P,LC-D4-13744.

1-038. Walpole-Westminister Bridge, Connecticut River, Walpole, Cheshire County, New Hampshire, 1910–1911 (demolished 1988). Joseph R. Worcester, engineer; Walsh's Holyoke Steam Boiler Works (Holyoke, MA), J. J. Fitzgerald (North Walpole, NH), contractors. Ernest Gould, photographer, 1988. P&P,HAER,NH,3-WALP.V,1-4.

1-039. Rio Puerco Bridge, Petrified Forest National Park, Holbrook Vic., Apache County, Arizona, 1931–1932 (demolished). U.S. Bureau of Public Roads, designer; W. E. Callahan Construction Company (Dallas, TX), contractor. Clayton B. Fraser, photographer, 1988. P&P,HAER,ARIZ,1-HOLB.V,1-3.

1-040. Detail of pier cap and cast steel rucker bearings, Rio Puerco Bridge, Petrified Forest National Park, Holbrook Vic., Apache County, Arizona, 1931–1932 (demolished). U.S. Bureau of Public Roads, designer; W. E. Callahan Construction Company (Dallas, TX), contractor. Clayton B. Fraser, photographer, 1988. P&P,HAER,ARIZ,1-HOLB.V,1-10.

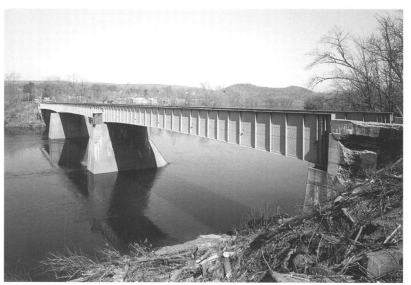

1-038

1-039

1-040

1-041. U.S. Route 1 Bridge, Wateree River, Camden Vic., Kershaw County, South Carolina, 1942. Vincennes Steel Corporation, builder. Wayne Moore, photographer, 1997. P&P,HAER,SC,28-CAMD.V,3-5.

This bridge combines continuous steel plate girders with a concrete T-beam deck structure.

1-042. Benton Street Bridge, Iowa River, Iowa City, Iowa, 1947–1949 (demolished, 1989). Edward L. Ashton, engineer; Jensen Construction Company, contractor. Bruce A. Harms, photographer, 1989. P&P,HAER,IOWA,52-IOWCI,4-4.

Edward Ashton (1903–1985) was a leading proponent of welded connections in bridge construction, which became increasingly common following World War II. This was the first all-welded bridge in Iowa.

1-043. Detail of welded connection, Benton Street Bridge, Iowa River, Iowa City, Iowa, 1947–1949 (demolished, 1989). Edward L. Ashton, engineer; Jensen Construction Company, contractor. Bruce A. Harms and Robert A. Ryan, photographers, 1989. P&P,HAER,IOWA,52-IOWCI,4-16.

The welded splice joining girder segments is just to the right of the bracket, which supports the sidewalk.

1-044. Construction of cantilevered plate deck girders, Hackensack Run Bridge II, New Jersey Turnpike, New Jersey, 1950–1952. Gottscho-Schleisner, photographer, 1951. P&P,LC-G613-60375.

1-041

1-042

1-044

1-043

CONCRETE GIRDER AND SLAB CONSTRUCTION

Reinforced concrete lends itself to a variety of configurations in beam construction. It can be cast in place or prefabricated and set into position by cranes. Increased strength and savings in the quantity of material may be obtained by pouring girders, decks, and even piers and abutments as monolithic structures or by prestressing components.

1-045

1-045. Asylum Avenue Viaduct, Second Creek, Knoxville, Tennessee, 1909 (demolished). W. B. Crenshaw and David Meriwether, engineers; Foster-Creighton-Gould Company (Nashville, TN), contractor. George Hornal, photographer, 1992. P&P,HAER, TENN,47-KNOVI,13-2.

This nine-span, 350-foot viaduct was one of the first continuous deck girder bridges erected in the United States.

1-046. View of deck girders, Asylum Avenue Viaduct, Second Creek, Knoxville, Tennessee, 1909 (demolished). W. B. Crenshaw and David Meriwether, engineers; Foster-Creighton-Gould Company (Nashville, TN), contractor. George Hornal, photographer, 1989. P&P,HAER,TENN,47-KNOVI,13-9.

1-047. Cement Plant Road Bridge, Leatherwood Creek, Bedford Vic., Lawrence County, Indiana, 1909. U.S. Cement Company, builder. Thomas W. Salmon II, Camille B. Fife, photographers, 1997. P&P,HAER,IND,47-BED.V,1-8.

This 73-foot, two-span, continuous deck girder bridge was built by a local cement company.

1-046

1-047

1-048. Fordway Bridge, Concord River, Billerica, Middlesex County, Massachusetts, 1912. J. R. Worcester, engineer; Charles R. Gow Company (West Roxbury, MA), contractor. Martin Stupich, photographer, 1996. P&P,HAER,MASS,9-BIL,8-6.

This bridge is an early example of T-beam construction in which the girders and deck are cast as a monolithic structure.

1-049. Detail of T-beams showing exposed steel reinforcement, Fordway Bridge, Concord River, Billerica, Middlesex County, Massachusetts, 1912. J. R. Worcester, engineer; Charles R. Gow Company (West Roxburgy, MA), contractor. Martin Stupich, photographer, 1996. P&P,HAER,MASS,9-BIL,8-16.

1-050. Papaahawahawa Gulch Bridge, Kipahula, Maui County, Hawaii, 1913, extended 1915. Maui County engineers, designers. David Franzen, photographer, 1996. P&P,HAER,HI,5-KIPLU,1-2.

The first span to be built, at right, is a reinforced concrete slab 11 feet long. The 22-foot span at left employs concrete girders.

1-048

1-049

1-050

1-051. Ravine Bluff Development Bridge (Sylvan Road Bridge), Glencoe, Cook County, Illinois, 1915 (replaced by a copy, 1985). Frank Lloyd Wright, designer. Thomas G. Yanul, photographer, 1984. P&P,HAER,ILL,16-GLENC,3-8.

The architect Frank Lloyd Wright (1867–1959) designed six houses and a number of other features including this reinforced concrete girder bridge in the Ravine Bluff subdivision developed by his close friend Sherman Booth. This and the following photographs were taken shortly before the bridge was replaced by a copy.

1-052. View of deck, Ravine Bluff Development Bridge (Sylvan Road Bridge), Glencoe, Cook County, Illinois, 1915 (replaced by a copy, 1985). Frank Lloyd Wright, designer. Thomas G. Yanul, photographer, 1984. P&P,HAER,ILL,16-GLENC,3-2.

1-053. View from below, Ravine Bluff Development Bridge (Sylvan Road Bridge), Glencoe, Cook County, Illinois, 1915 (replaced by a copy, 1985). Frank Lloyd Wright, designer. Thomas G. Yanul, photographer, 1984. P&P,HAER,ILL,16-GLENC,3-12.

1-054. San Antonio Creek Bridge, Lompoc
Vic., Santa Barbara County, California,
1916. Mayberry & Parker (Los Angeles, CA),
engineers. Ed Andersen, photographer,
1982. P&P,HAER,CAL,42-LOMP.V,3-5.

1-054

1-055. Detail of transverse arch-rib deck
structure, San Antonio Creek Bridge, Lompoc Vic., Santa Barbara County, California,
1916. Mayberry & Parker (Los Angeles, CA),
engineers. Ed Andersen, photographer,
1982. P&P,HAER,CAL,42-LOMP.V,3-7.

Edward L. Mayberry Jr., a prominent California engineer specializing in steel and concrete construction, held a patent for the
system of transverse arches that support
the deck instead of the more common
arrangement of floor beams and stringers
(see 1-060).

1-055

1-056. Main Street Bridge, Fox River, West
Dundee, Kane County, Illinois, 1917–1918.
George N. Lamb, engineer; Oltendorf Con-
struction Company (Palatine, IL), contractor.
Michael A. Dixon, photographer, 1994. P&P,
HAER,ILL,45-WEDUN.V,1-2.

Each 60-foot span consists of six T-beams
that were poured in place. Their inverted
arch shape is unusual.

1-056

1-057

1-058

1-059

1-060

1-057. Big Conestoga Creek Bridge No. 12, Brownstown Vic., Lancaster County, Pennsylvania, 1917. Frank H. Shaw, county engineer, designer; Paul D. Kauffman (Reading, PA), contractor. Jet T. Lowe, photographer, 1999. P&P,HAER,PA,36-BROTO.V,1-5.

This 122-foot bridge incorporates girder, arch, and cantilevered construction.

1-058. Salt River Bridge, Ferndale Vic., Humboldt County, California, 1919. Henry John Brunnier (San Francisco, CA), engineer; Thomas Engelhart (Eureka, CA), contractor. Ed Andersen, photographer, 1992. P&P,HAER,CAL,12-FERDA.V,1-7.

At the time of completion, the 142-foot reinforced concrete girder spans used in this bridge were the world's longest. The long girders are 12 feet deep. The designer, Henry John Brunnier (1882–1971), one of the foremost structural engineers in San Francisco, was an innovator in reinforced concrete construction.

1-059. Deck, Salt River Bridge, Ferndale Vic., Humboldt County, California, 1919. Henry John Brunnier (San Francisco, CA), engineer; Thomas Engelhart (Eureka, CA), contractor. Ed Andersen, photographer, 1992. P&P,HAER,CAL,12-FERDA.V,1-6.

1-060. Understructure showing girders, floor beams, and stringers, Salt River Bridge, Ferndale Vic., Humboldt County, California, 1919. Henry John Brunnier (San Francisco, CA), engineer; Thomas Engelhart (Eureka, CA), contractor. Ed Andersen, photographer, 1992. P&P,HAER,CAL,12-FERDA.V,1-10.

1-061. Black River Bridge, Neillsville, Clark County, Wisconsin, 1921. J. H. A. Brahtz, engineer. John N. Vogel, photographer, 1994. P&P,HAER,WIS,010-NEIL,1-3.

Like the Salt River Bridge, this is a deep beam structure. The mass of the girders is greatest at the areas subject to the highest bending stresses.

1-062. Puente San Antonio, San Antonio Channel, San Juan, Puerto Rico, 1924–1925. Rafael Nones, engineer; Robert Prann, engineer, contractor. Héctor Méndez-Cantini, photographer, 1995. P&P,HAER,PR,7-SAJU,62-6.

Behind the ornamental arches of the exterior faces is a conventional structure of reinforced concrete girders.

1-063. Winnebago River Bridge, Mason City Vic., Cerro Gordo County, Iowa, 1926. Iowa State Highway Commission, designer; William Henkel (Mason City, IA), contractor. Bruce A. Harms, photographer, 1996. P&P,HAER,IOWA,17-MASCIT,10-7.

The bridge has three spans (the center span is 70 feet) consisting of four monolithic cantilevered girders. It is part of a family of reinforced concrete highway bridges introduced in Iowa in 1917.

1-064

1-064. Monroe Street Bridge, River Raisin, Monroe, Monroe County, Michigan, 1927–1929 (demolished). Michigan State Highway Department, designer; W. H. Knapp Company, contractor. Carla Anderson, photographer. P&P,HAER,MICH,58-MONRO,3-7.

1-065. Ducktrap Bridge, Duck Trap River, Lincolnville, Waldo County, Maine, 1919–1920 (lower bridge), 1932–1933 (upper bridge); replaced 2000. Lower bridge: L. B. Jones, engineer; Cyr Brothers (Waterville, ME), contractor. Brian Vandenbrink, photographer, 1997. P&P,HAER,ME,14-LINC,2-12.

In 1932, the highway through Lincolnville was designated part of U.S. Route 1. Engineers eliminated the steep, traffic-slowing approaches to the Ducktrap River crossing by superimposing a high bridge on the older bridge. The unusual structure became a local landmark. It was replaced by a new bridge of similar appearance in 2000.

1-065

1-066. Fifth Street viaduct, spanning Bacon's Quarter Branch Valley, Richmond, Virginia, 1933. Alfredo C. Janni, engineer; Richmond Bridge Corporation, contractor. Robert C. Shelley, photographer, 1992. P&P,HAER,VA,44-RICH,115-8.

This is an example of rigid frame construction in which the supports, girders, and deck are formed as a monolithic, continuous-span structure. The technique was introduced to the United States from Germany and Brazil in the 1920s by Arthur G. Hayden.

1-067. Norris Freeway Bridge, Norris Vic., Anderson County, Tennessee, ca. 1936–1942. Roland Wank, chief architect, Tennessee Valley Authority, designer. FSA/Office of War Information photograph, between 1934 and 1945. P&P,LC-USW33-015670-ZC.

1-068. White Oak Shade Road Bridge, Merritt Parkway, New Canaan, Fairfield County, Connecticut, 1937. George L. Dunkelberger, head architect, Connecticut Highway Department, designer; M. A. Gammino Construction Company (Providence, RI), contractor. Jet T. Lowe, photographer, 1992. P&P,HAER,CONN,1-NECA,11-1.

Like many subsequent reinforced concrete highway overpasses, this is a rigid frame structure. The structural system allowed Dunkelberger to reduce the thickness of the span at the center, creating the arched effect that, along with the fine proportions and elegant detailing, adds to the bridge's expressive character.

1-066

1-067

1-068

1-069

1-069. Cloverleaf Interchange and Bridge No. 5820, Inver Grove Heights, Dakota County, Minnesota, 1940. E. J. Miller, engineer; Minnesota Department of Highways, designer; Anderson & Sons, contractor. Mike Whyte, photographer, 1994. P&P, HAER,MINN,19-IVGRHE,1A-1.

This is an early example of a cloverleaf interchange in Minnesota.

1-070. Bridge No. 5820, Cloverleaf Interchange, Inver Grove Heights, Dakota County, Minnesota, 1940. E. J. Miller, engineer; Minnesota Department of Highways, designer; Anderson & Sons, contractor. Mike Whyte, photographer, 1994. P&P,HAER,MINN,19-IVGRHE,1A-5.

1-071. View of continuous deck girders, Bridge No. 5820, Cloverleaf Interchange, Inver Grove Heights, Dakota County, Minnesota, 1940. E. J. Miller, engineer; Minnesota Department of Highways, designer; Anderson & Sons, contractor. Mike Whyte, photographer, 1994. P&P,HAER,MINN,19-IVGRHE,1A-6.

1-070

1-071

1-072. Walnut Lane Bridge, Monoshone Creek, Philadelphia, Pennsylvania, 1949–1950 (superstructure demolished and reconstructed 1989–1990). Gustav Magnel (Ghent, Belgium), engineer; Preload Corporation (New York City, NY), beam fabricator; Henry Horst Company (Philadelphia, PA), contractor. A. Pierce Bounds, photographer, 1988. P&P,HAER,PA,51-PHILA,715-5.

This reinforced concrete I-beam bridge has three 74-foot simple spans. It was the first prestressed concrete beam bridge in the United States and served as an important prototype in the development of prefabricated, prestressed concrete highway bridges built throughout the country in the second half of the 20th century.

1-073. Linn Cove Viaduct, Grandfather Mountain, Blue Ridge Parkway, Avery County, North Carolina, 1983. Figg Engineering Group, designer; Jasper Construction, builder; construction, 1983. Hugh Morton, photographer, 2006. © Grandfather Mountain, 1983; 2006. P&P,LC-DIG-ppem-00060.

The last component of the Blue Ridge Parkway to be completed, the 1,243-foot viaduct wraps across the face of Grandfather Mountain forming an S-curve. The 180-foot spans consist of cantilevered, prefabricated concrete segments integrating box girders (that is, shaped to form a rectangular tube) and deck. To minimize damage to the environmentally sensitive site, the bridge was built from above by trucking the segments along the completed portions and dropping them into place with a crane mounted on the roadway. Black iron oxide was added to the concrete to give it a color complementing the mountain's rocky outcrops.

1-074. Construction photo, Linn Cove Viaduct, Grandfather Mountain, Blue Ridge Parkway, Avery County, North Carolina, 1983. Figg Engineering Group, designer; Jasper Construction, builder; construction, 1983. Hugh Morton, photographer, 2006. © Grandfather Mountain, 1983; 2006. P&P,LC-DIG-ppem-00059.

1-072

1-073

1-074

ARCH BRIDGES

AN ARCH DISTRIBUTES ITS LOAD to its points of support radially. This property

makes it possible to construct arches with small components—such as blocks

of stone, bricks, or short lengths of wood or metal—that are wedged together

by the radial thrust, which is resisted at the ends of the span by the abutments.

An ancient structural form, the arch has been used throughout the history of

American bridge building in masonry, metal, wood, and concrete construction.

2-001

2-002

2-003

Masonry Road Bridges

2-001. Pennypack Creek Bridge, Philadelphia, Pennsylvania, ca. 1697. Joseph Elliott, photographer, 1997. P&P,HAER,PA,51-PHILA,710-1.

Though significantly altered by widening, changes to the roadbed, and reinforcement, this three-arch stone structure (maximum span 25 feet) built on the colonial King's Highway is believed to be the oldest extant bridge in the United States. Its dimensions and construction techniques reflect traditional European practice.

2-002. Choate Bridge, Ipswich River, Ipswich, Massachusetts, 1764 (widened 1838). John Choate, designer. Jet T. Lowe, photographer, 1986. P&P,HAER,MASS,5-IPSWI,8-14.

John Choate was a prominent Ipswich attorney who also had expertise in stone masonry.

2-003. Skippack Stone Arch Bridge, Skippack Creek, Skippack, Montgomery County, Pennsylvania, 1792. Stephen Lane, John Alman, John Burke, masons. Joseph Elliott, photographer, 1994. P&P,HAER,PA,46-SKIPP,1-3.

This bridge of eight 20-foot sandstone arches was built as part of a road improvement campaign. It provided a focal point, attracting innkeepers and other businessmen to the area.

2-004. Casselman River Bridge, Grantsville Vic., Garrett County, Maryland, 1813. David Shriver Jr., builder. A. S. Burns, photographer, 1933. P&P,HABS,MD,12-GRANT.V,1-2.

Built to accommodate increasing traffic on the National Road, the bridge has an 80-foot span, which, at the time of construction, made it the longest masonry-arch bridge in the United States.

2-005. Wilson Bridge, Conococheague Creek, Hagerstown Vic., Washington County, Maryland, 1817–1819 (restored 1984). Silas Harry, builder. William Edmund Barrett, photographer, 1982. P&P,HAER,MD,22-HAGTO.V,21.

This five-span masonry bridge of local limestone has dressed voussoirs and a rubble core. It was erected by the Hagerstown and Conococheague Turnpike Company as part of the National Road. The builder, Silas Harry, also constructed bridges in Pennsylvania and West Virginia.

2-006. Elevations and sections, Wilson Bridge, Conococheague Creek, Hagerstown Vic., Washington County, Maryland, 1817–1819. Silas Harry, builder. Washington County Engineering Department, delineator, 1982. P&P,HAER,MD,22-HAGTO.V-, sheet no. 2.

2-004

2-005

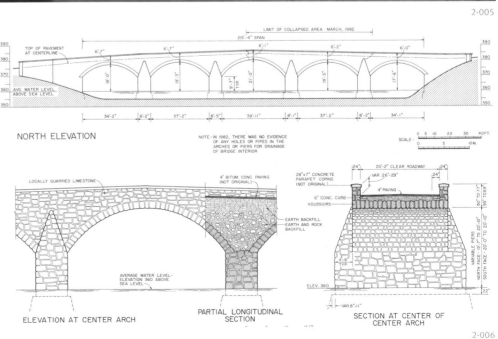

2-006

2-007

2-007. Poinsett Bridge, Little Gap Creek, Tigerville Vic., Greenville County, South Carolina, completed 1820. Design attributed to Robert Mills. Jack E. Boucher, photographer, 1988. P&P,HAER, SC,23-TIGVI.V,1-10.

This is the oldest bridge in South Carolina. The South Carolina Board of Public Works commissioned it as part of a state highway linking Charleston, Greenville, and Asheville, North Carolina. The builders spanned the creek with an elegantly framed pointed arch that transmits its radial thrust in a more downward direction than round arches. The design has been attributed to Robert Mills (1781–1855), regarded as the first professionally trained architect born in the United States.

2-008. View of roadway, Poinsett Bridge, Little Gap Creek, Tigerville Vic., Greenville County, South Carolina, completed 1820. Design attributed to Robert Mills. Jack E. Boucher, photographer, 1988. P&P,HAER,SC,23-TIGVI.V,1-3.

2-009. East Bridge, Maclay's Mill Twin Bridges, Conodoguinet Creek, Mowersville Vic., Franklin County, Pennsylvania, 1827. Silas Harry, builder. Joseph Elliott, photographer, 1997. P&P, HAER,PA,28-MOWVI.V,1-2.

Silas Harry, a stone mason, designed and built at least five bridges in Franklin County and others elsewhere in Pennsylvania, Maryland, and West Virginia (2-007). This structure, consisting of three spans of coursed limestone blocks and rubble, was built at the site of a mill and highway junction; it leads to a similar two-span bridge that completes the stream crossing.

2-008

2-009

2-010

2-011

2-010. Main Street Bridge, Genesee River, Rochester, New York, 1855–1857. Kaufman & Bissel, engineers; I. F. Quinby, consulting engineer. Martin Stupich, photographer, 1987. P&P,HAER,NY,28-ROCH,43-2.

This is the fourth bridge to occupy the site since 1812. Like the wooden structure it replaced, it had a row of market buildings supported on piers alongside it until the mid-1960s. Kaufman & Bissel was an engineering firm; and its involvement exemplifies how by the mid-nineteenth century professional engineers increasingly assumed responsibility for designing bridges, especially in urban areas.

2-011. Main Street Bridge, Genesee River, Rochester, New York. Graphite on wove paper. Signed S. S. C., dated January 10, 1855. P&P,DRWG/US—no. 39 (A size).

This drawing shows the bridge before construction of the present stone bridge. Extensions on the north side of the piers supported market buildings. Though the practice of erecting houses and shops on bridges was once commonplace in Europe, it had fallen out of favor by the nineteenth century, and its appearance in Rochester is most unusual. High real estate prices on land and the prospect of a steady parade of customers must have offset the considerable risks of fire and flood in the minds of the builders.

2-012

2-012. Jefferson Street Bridge, East Branch of Wears Creek, Jefferson City, Missouri, 1857 (deck modified 1964, 1980). Dr. William Armstrong Davison Sr., contractor. Lloyd Grotjan, photographer, 1987. P&P,HAER,MO,26-JEFCI,16-4.

2-013. Elkader Bridge, Turkey River, Elkader, Clayton County, Iowa, 1888–1889. M. Tschirgi Jr. (Dubuque, IA), engineer; Byrne & Blake, (Dubuque, IA), contractor. Joseph Elliott, photographer, 1995. P&P,HAER,IOWA,22-ELKA,1-3.

This bridge replaced an iron truss structure that had been a maintenance nightmare. The availability of local stone and a skilled labor force made it possible to build a masonry structure at a lower cost than an iron bridge with similar capacity.

2-013

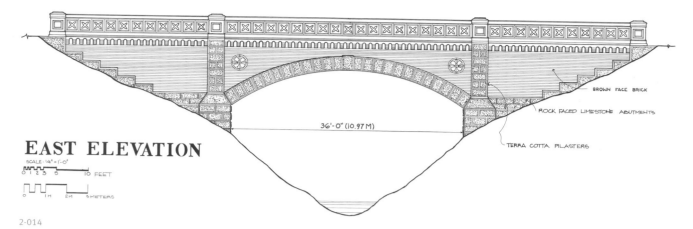

EAST ELEVATION

SCALE: ¼" = 1'-0"

0 1 2 3   5   10 FEET

0   1 M   2 M   3 METERS

36'-0" (10.97 M)

BROWN FACE BRICK

ROCK FACED LIMESTONE ABUTMENTS

TERRA COTTA PILASTERS

2-014

2-015

2-014. Elevation, Lake Park Brick Arch Bridge, Milwaukee, Wisconsin, 1893. Oscar Sanne, engineer; Gerhard F. Stuewe Company, contractor. Scott Barber, delineator, 1987. P&P,HAER,WIS,40-MILWA,49-,sheet no. 1.

The ornamental Brick Arch Bridge marks the northern entrance of Lake Park, designed by landscape architect Frederick Law Olmsted Sr. The five layers of brick forming the arch are faced with terra cotta panels shaped like rusticated stone voussoirs. The roadbed rests on a fill of concrete and aggregate framed by brick.

2-015. Detail of pier and arch, Lake Park Brick Arch Bridge, Milwaukee, Wisconsin, 1893. Oscar Sanne, engineer; Gerhard F. Stuewe Company, contractor. Martin Stupich, photographer, 1987. P&P,HAER,WIS,40-MILWA,49-3.

2-016. Mill Bridge, Little Wolf River, Scandinavia, Waupaca County, Wisconsin, 1907. Lars Anderson, local mason, designer and builder. Martin Stupich, photographer, 1987. P&P,HAER,WIS,68-SCAN,1-3.

Ring arches (the arches defining the shape of the arch) of cut sandstone frame this structure of similarly sized (around 10 to 16 inches in diameter) granite fieldstones set in mortar. This construction exemplifies a regional building practice based on readily available materials and local expertise.

2-017. Wrightstown Bridge, East River, Wrightstown, Brown County, Wisconsin, 1909. Martin Stupich, photographer, 1987. P&P,HAER,WIS,5-WRITO,1-1.

2-016

2-017

2-018. Pine Creek Bridge, Zion–Mount Carmel Highway, Springdale Vic., Washington County, Utah, 1930. B. J. Pinch, engineer, U.S. Department of Agriculture, Bureau of Public Roads; Thomas C. Vint, landscape architect, U.S. Department of the Interior, National Park Service. Clayton B. Fraser, photographer, 1984. P&P,HAER, UTAH,27-SPDA.V,3B-4.

2-019. Cutaway view, Pine Creek Bridge, Zion–Mount Carmel Highway, Springdale Vic., Washington County, Utah, 1930. B. J. Pinch, engineer, U.S. Department of Agriculture, Bureau of Public roads; Thomas C. Vint, landscape architect, U.S. Department of the Interior, National Park Service. Todd A. Croteau, delineator, 1993. P&P,HAER, UTAH,27-SPDA.V,3B-,sheet no. 2.

The Pine Creek Bridge has concrete foundations and spandrel walls, but the building of its 60-foot native sandstone arch employed techniques that have been used for thousands of years. During construction, wooden centering (also called falsework) defined the shape of the arch and supported the voussoirs (wedge-shaped blocks), which form the arch rings and frame a core of coursed rubble set in concrete. The choice of masonry was an aesthetic decision intended to harmonize with the rocky character of the site and the nearby Great Zion Arch downstream.

2-018

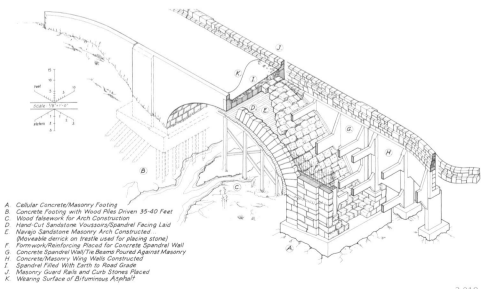

A. Cellular Concrete/Masonry Footing
B. Concrete Footing with Wood Piles Driven 35-40 Feet
C. Wood falsework for Arch Construction
D. Hand-Cut Sandstone Voussoirs/Spandrel Facing Laid
E. Navajo Sandstone Masonry Arch Constructed...
     (Moveable derrick on trestle used for placing stone)
F. Formwork/Reinforcing Placed for Concrete Spandrel Wall
G. Concrete Spandrel Wall/Tie Beams Poured Against Masonry
H. Concrete/Masonry Wing Walls Constructed
I. Spandrel Filled With Earth to Road Grade
J. Masonry Guard Rails and Curb Stones Placed
K. Wearing Surface of Bituminous Asphalt

2-019

Masonry Aqueducts

Aqueducts most commonly carry water for drinking or irrigation, but some are built to carry canals across obstacles such as rivers and ravines.

2-020

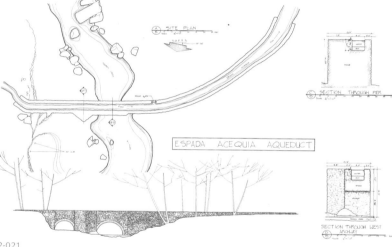

2-021

2-022

2-020. Espada Acequia, La Piedra Creek, San Antonio, Texas, ca. 1731. Jet T. Lowe, photographer, 1983. P&P,HAER,TEX,15-SANT.V,4A-1.

2-021. Plan and elevation, Espada Acequia, La Piedra Creek, San Antonio, Texas, ca. 1731. Gary Rogers, delineator, 1973. P&P, HAER,TEX,15-SANT.V,4A-,sheet no. 1.

The La Piedra aqueduct is part of a still-functioning irrigation (acequia) system built by Franciscan missionaries and Pacaos Indians as part of the San Francisco de la Espada mission.

2-022. High Bridge, Old Croton Aqueduct, Harlem River, New York, New York, 1839–1848. John B. Jervis, engineer; James Renwick Jr., architect. DETR, photographer, ca. 1900. P&P,LC-USZC4-6323.

The High Bridge originally supported two 33-inch-diameter cast-iron pipes conveying drinking water into Manhattan as part of the 40-mile Croton Aqueduct. The monumental structure celebrated the aqueduct as a symbol of progress. In 1923, a steel span replaced the five 80-foot arches crossing the navigable channel.

2-023. Cabin John Aqueduct Bridge, Cabin John Creek, Cabin John, Mont-gomery County, Maryland, 1853–1864. Montgomery C. Meigs and Alfred L. Rives, engineers. Jet T. Lowe, photographer, 1985. P&P,HAER,MD,16-CABJO, 1-2.

At the time of its completion, the 220-foot stone-masonry arch was the longest in the world, and it remains the longest in the United States. It contin-ues to carry water to the District of Columbia as part of the Washington Aque-duct system.

2-024. Construction photo of centering, Cabin John Aqueduct Bridge, Cabin John Creek, Cabin John, Montgomery County, Maryland, 1853–1864. Mont-gomery C. Meigs and Alfred L. Rives, engineers. Unidentified photographer, 1859. P&P,LC-USZ6-1269.

2-025. Echo Bridge, Sudbury Aqueduct, Charles River, Newton, Massachus-sets, 1876. Alphonse Fteley, chief engineer; George W. Phelps (Springfield, MA), contractor. Jet T. Lowe, photographer, 1982. P&P,HAER,MASS,9-NEWT,15-2.

The bridge carries drinking water diverted from the Sudbury River by the Boston Water Company. It has a concrete core (an early use of the material) contained by granite rubble arches with ashlar (squared stone) facing.

2-026

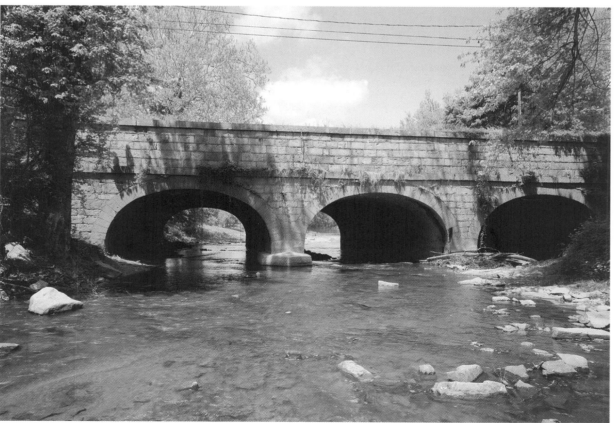

2-027

2-026. Monocacy Aqueduct, Chesapeake & Ohio Canal, Monocacy River, Dickerson Vic., Frederick County, Maryland, 1829–1833 (rehabilitated 2005). Benjamin Wright, engineer. William Henry Jackson, photographer, ca. 1892. P&P,LC-D43-1776.

Seven granite arches, each spanning 54 feet, carried the canal and its towpath across the Monocacy River as part of a waterway between Georgetown, District of Columbia, and Cumberland, Maryland.

2-027. Wiconisco Canal Aqueduct No. 3, Powell Creek, Halifax Vic., Dauphin County, PA, 1838–1840. John P. Bailey, William R. Harrison, engineers; George Mish & Company, builders. Jet T. Lowe, photographer, 1999. P&P,HAER,PA,22-HAFX.V,1-2.

2-028

2-028. Schoharie Creek Aqueduct, enlarged Erie Canal, Fort Hunter, Montgomery County, New York, 1839–1841. John B. Jervis, engineer. Jack E. Boucher, photographer, 1969. P&P,HAER,NY,29-FORHU,2A-26.

Before the construction of the aqueduct, barge traffic had to leave the canal channel and cross the creek, which sometimes was unnavigable. The arches carried the towpath alongside the waterway, which was contained in a tank constructed of tightly fitted wooden planks supported by the masonry piers visible at the left.

2-029. Site map, Schoharie Creek Aqueduct, enlarged Erie Canal, Fort Hunter, Montgomery County, New York, 1839–1841. John B. Jervis, engineer. David Bouse, delineator. P&P,HAER,NY,29-FORHU,2A-,sheet no. 2.

2-030. Section and elevation, Schoharie Creek Aqueduct, enlarged Erie Canal, Fort Hunter, Montgomery County, New York, 1839–1841. John B. Jervis, engineer. David Bouse, delineator. P&P,HAER,NY,29-FORHU,2A-,sheet no. 3.

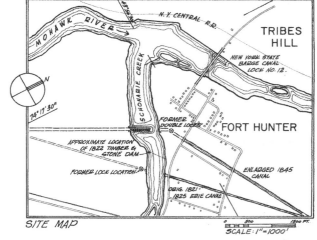

SITE MAP

2-029

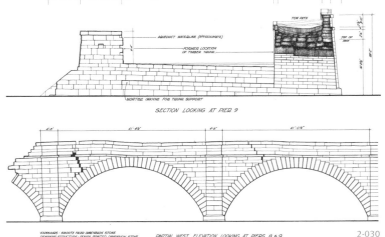

SECTION LOOKING AT PIER 9

PARTIAL WEST ELEVATION LOOKING AT PIERS 8 & 9

2-030

Masonry Railroad Bridges

The high compressive strength and stability of masonry arches make them well suited for railroad bridges, and in the early days of the railroad industry, their monumental appearance enhanced the image of prosperity and permanence sought by ambitious companies, such as the Baltimore & Ohio Railroad. These advantages, however, were offset by a number of drawbacks, including the relatively long time required for construction, the cost of transporting building stone over long distances when not available on site, and practical limitations to the length of unsupported spans. Iron and steel became the materials of choice in the mid-nineteenth century, but following a number of structural failures at the century's end, some railroad companies, such as the Pennsylvania, returned to masonry construction for bridges on trunk lines.

2-031

2-031. Thomas Viaduct, Baltimore & Ohio Railroad, Patapsco River, Elridge Landing and Relay, Howard and Baltimore Counties, Maryland, 1833–1835. Benjamin H. Latrobe II, engineer; John McCartney, builder. Engraving in Charles A. Dana, ed., The United States Illustrated, ca. 1858. P&P,HAER,MD,14-ELK,1-17.

Fully 612 feet long with arches spanning about 58 feet, this granite bridge, built on a curved alignment, was the first multi-span masonry railroad bridge in the United States. It remains in use today. Benjamin Henry Latrobe II (1806–1878) was the son of the prominent architect and engineer Benjamin Henry Latrobe (1764–1820), who pioneered American recognition of both fields as professional disciplines. Trained as an attorney, the younger Latrobe turned to engineering in 1829. He began his career with the B&O, measuring stone for track ballast and worked his way up through the ranks to the position of chief engineer of the rapidly expanding railroad.

2-032. Thomas Viaduct, Baltimore & Ohio Railroad, Patapsco River, Elridge Landing and Relay, Howard and Baltimore Counties, Maryland, 1833–1835. Benjamin H. Latrobe II, engineer; John McCartney, builder. William Edmund Barrett, photographer, 1970. P&P,HAER,MD,14-ELK,1-8.

2-032

2-033. Cast-iron railing, Thomas Viaduct, Baltimore & Ohio Railroad, Patapsco River, Elridge Landing and Relay, Howard and Baltimore Counties, Maryland, 1833–1835. Benjamin H. Latrobe II, engineer; John McCartney, builder. William Edmund Barrett, photographer, 1970. P&P,HAER,MD,14-ELK,1-10.

2-034. Canton Viaduct, Boston & Providence Railroad, Neponset River, Canton, Norfolk County, Massachusetts, 1835. William Gibbs McNeil, engineer. Peter Stott, photographer, ca. 1984. P&P, HAER,MASS,11-CANT,2-1.

2-035

2-036

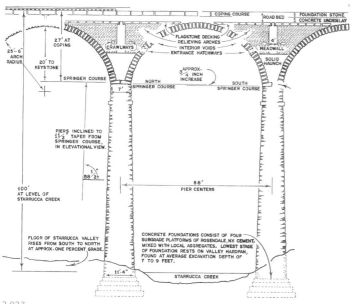

2-037

2-035. Aerial view, Starrucca Viaduct, New York and Erie Railroad, Starrucca Creek, Landesboro, Susquehanna County, Pennsylvania, 1848. James P. Kirkwood and Julius W. Adams, engineers. Jack E. Boucher, photographer, 1971. P&P,HAER, PA,58-LANBO,1-15.

Seventeen arches of local bluestone (a type of sandstone), 110 feet high with 50-foot spans, carry the railroad across the broad valley formed by Starrucca Creek.

2-036. Starrucca Viaduct, New York and Erie Railroad, Starrucca Creek, Landesboro, Susquehanna County, Pennsylvania, 1848. James P. Kirkwood and Julius W. Adams, engineers. Jack E. Boucher, photographer, 1971. P&P,HAER, PA,58-LANBO,1-13.

2-037. Elevation, Starrucca Viaduct, New York and Erie Railroad, Starrucca Creek, Landesboro, Susquehanna County, Pennsylvania, 1848. James P. Kirkwood and Julius W. Adams, engineers. Paul Berry, delineator, 1984. P&P,HAER, PA,58-LANBO,1-,sheet no. 3.

2-038. Peacock's Lock Viaduct, Philadelphia & Reading Railroad, Schuylkill River, Reading Vic., Berks County, Pennsylvania, 1853–1856. Gustavus A. Nicolls, engineer. William Edmund Barrett, photographer, 1975. P&P,HAER,PA,6-READ.V,2-2.

The circular openings in the spandrels (the triangular areas between the arches and the deck) are unusual in American masonry bridges, but they illustrate a typical situation in arched construction. Since the primary structural loads are accommodated within the depth of the arch rings, much of the material in the spandrels is infill, which designers often seek to minimize. The spaces behind the spandrels of many masonry briges are hollow or filled with something lighter than the stone forming the arch rings. The designers of reinforced concrete bridges often leave the spandrels open.

2-039. Plan of Savannah Repair Yards, 1860 Brick Arch Viaduct, Central of Georgia Railway, Ogeechee Canal, Savannah, Georgia, 1860. Martin Mueller and Augustus Schwaab, architecture and engineering. Roland David Schaaf, delineator. P&P,HAER,GA,26-SAV,55-,sheet no. 2.

The Savannah Yards, begun in 1853, were a comprehensive facility incorporating passenger, freight, repair, and administrative functions. The 1860 viaduct served the freight facilities (upper part of the plan).

2-040. 1860 Brick Arch Viaduct, Central of Georgia Railway, Ogeechee Canal, Savannah, Georgia, 1860. Martin Mueller and Augustus Schwaab, architecture and engineering. Jack E. Boucher, photographer, 1976. P&P,HAER,GA,26-SAV,18-1.

Large quantities of building stone were not readily available in Savannah, but the city had an efficient brick industry that supplied the Savannah gray brick used throughout the Central of Georgia yards.

2-038

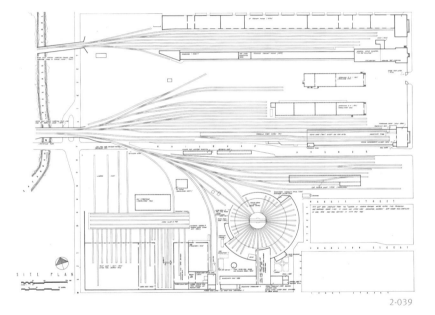

2-039

2-040

2-041

2-042

2-041. Chicago & North Western Railway Bridge No. 128, Turtle Creek, Tiffany, Rock County, Wisconsin, 1869. Van Mienan, chief engineer; John Watson, builder. Martin Stupich, photographer, 1987. P&P,HAER,WIS,53-TIF,1-5.

2-042. Stone Arch Bridge, Minneapolis Union Railway Company (Great Northern Railroad), Mississippi River, Minneapolis, Minnesota, 1883. Charles C. Smith, engineer. DETR, photographer, ca. 1900–1910. P&P,LC-D4-36332.

2-043. Rockville Bridge, Pennsylvania Railroad, Susquehanna River, Rockville, Dauphin County, Pennsylvania, 1900–1902. William H. Brown, chief engineer; George Nauman and H. S. Righter, assistant engineers; Drake & Stratton Company, H. S. Kerbaugh Inc., contractors. R. W. Johnston, photographer, ca. 1905. P&P, PAN US GEOG—Pennsylvania, no. 20 (F size).

A product of the Pennsylvania Railroad's campaign to replace iron briges with more durable masonry structures, the Rockville Bridge consists of forty-eight 70-foot arches and has an total length of 3,791 feet, making it the longest stone arch railway bridge in the world. The arch rings and facing walls are sandstone, and the interior cavities are filled with concrete.

2-043

Iron and Steel Arch Bridges

The oldest surviving iron bridge is an arch composed of cast-iron ribs built across the Severn River near Coalbrookdale, England, in 1779. An early promoter of the use of iron for bridges was the Anglo-American political theorist Thomas Paine (1737–1809), who patented a design for a cast-iron arch bridge in England in 1788. The arch form, which places all members in compression, was well suited to the structural properties of cast iron, which is strong in compression but relatively weak in tension. Since the 1870s, the inherent structural efficiency of the arch has also been applied to steel bridges. Arches can support decks from above or below.

2-044

2-044. Dunlap's Creek Bridge, Brownsville, Fayette County, Pennsylvania, 1836–1839. Richard Delafield, engineer; Keys & Searight, contractor; Herbertson Foundry, John Snowden Company (Brownsville, PA), cast- and wrought-iron fabrication. John Kennedy Lacock, photographer, 1910. P&P,LC-USZ62-46243.

A lack of high-quality building stone, a desire for a more durable structure than a wooden span, and the presence of skilled ironworkers inspired Captain Richard Delafield (1798–1873) of the U.S. Army Corps of Engineers to design the nation's first all-metal bridge in Brownsville as part of the National Road.

2-045. Detail of original arch with modern bracing and roadbed extension, Dunlap's Creek Bridge, Brownsville, Fayette County, Pennsylvania, 1836–1839. Richard Delafield, engineer; Keys & Searight, contractor; Herbertson Foundry, John Snowden Company (Brownsville, PA), cast- and wrought-iron fabrication. Jet T. Lowe, photographer, 1983. P&P,HAER,PA,26-BROVI,2-4.

2-045

2-046. Isometric view, Dunlap's Creek Bridge, Brownsville, Fayette County, Pennsylvania, 1836–1839. Richard Delafield, engineer; Keys & Searight, contractor; Herbertson Foundry, John Snowden Company (Brownsville, PA), cast- and wrought-iron fabrication. Christopher H. Marston, delineator, 1992. P&P,HAER,PA,26-BROVI, 2-,sheet no. 3.

Cast-iron tubes, bolted together, form the 80-foot arch, which supports a system of iron trusses and plates that underlie the roadway, originally macadam.

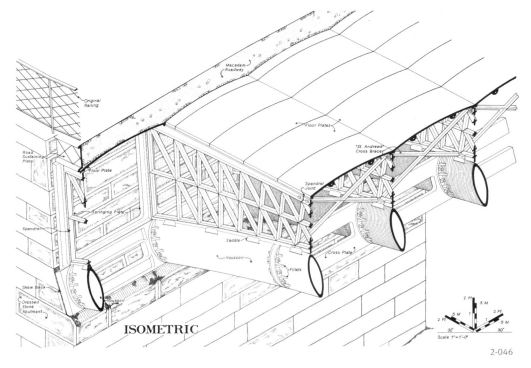

ISOMETRIC

2-046

2-047

2-048

2-047. Gothic Arch (Bridge No. 28), Central Park, New York, New York, 1864. Calvert Vaux, designer. Jet T. Lowe, photographer, 1984. P&P,HAER, NY,31-NEYO,153C-3.

This and the bridges shown in 2-048 and 2-049 have more pronounced characteristics of arches than Calvert Vaux's two other iron bridges in the park (see 1-019, 1-020).

2-048. Central Park Bridge No. 27, Central Park, New York, New York, 1864. Calvert Vaux, designer. Jet T. Lowe, photographer, 1984. P&P,HAER,NY,31-NEYO,153D-1.

2-049. Bridge No. 24, Central Park, New York, New York, 1865. Calvert Vaux, designer. Jet T. Lowe, photographer, 1984. P&P,HAER, NY,31-NEYO,153E-1.

2-049

2-050

2-050. Eads Bridge, Mississippi River, St. Louis, Missouri, 1867–1874. James B. Eads, engineer-in-chief; James Andrews (Pittsburgh, PA), masonry contractor; Keystone Bridge Company (Pittsburgh, PA), prime contractor for iron and steel. Color lithograph published by Compton & Company (St. Louis, MO), 1874. P&P,LC-USZC4-4899.

With separate decks for road and railroad traffic, the privately funded Eads Bridge was the first Mississippi River crossing at St. Louis. It ranks among the foremost engineering works of the nineteenth century for the unprecedented use of steel at such a scale; the use of cantilevered construction in place of centering, which would have obstructed navigation; and the use of pneumatic caissons to excavate the foundations for the piers. Its completion was a major civic event commemorated by the publication of souvenir lithographs, such as this, which surrounds a view of the bridge with a portrait of James Eads and construction scenes based on photographs taken by R. Benecke.

2-051. "The Great Tornado at St. Louis, Mo. and E. St. Louis Ill's, May 27th 1896." Commemorative lithograph published by Kurz & Allison, June 18, 1896. P&P,LC-USZ62-526.

The Eads Bridge is depicted as the only structure unscathed by the tornado, which killed approximately 255 people, injured more than a thousand, and caused great property damage on both sides of the Mississippi River.

2-051

2-052

2-053

2-054

2-055

2-052. Erection of western arch showing cable stays supporting cantilevers, Eads Bridge, Mississippi River, St. Louis, Missouri, 1867–1874. James B. Eads, engineer-in-chief; James Andrews (Pittsburgh, PA), masonry contractor; Keystone Bridge Company (Pittsburgh, PA), prime contractor for iron and steel. C. M. Woodward, History of the St. Louis Bridge (1881), plate 43. P&P,LC-USZ62-71349.

2-053. Erection of highway deck on the western arch, Eads Bridge, Mississippi River, St. Louis, Missouri, 1867–1874. James B. Eads, engineer-in-chief; James Andrews (Pittsburgh, PA), masonry contractor; Keystone Bridge Company (Pittsburgh, PA), prime contractor for iron and steel. C. M. Woodward, History of the St. Louis Bridge (1881), plate 45. P&P,LC-USZ62-69757.

2-054. Eads Bridge, Mississippi River, St. Louis, Missouri, 1867–1874. James B. Eads, engineer-in-chief; James Andrews (Pittsburgh, PA), masonry contractor; Keystone Bridge Company (Pittsburgh, PA), prime contractor for iron and steel. Jet T. Lowe, photographer, 1983. P&P,HAER,MO,96-SALU,77-3.

2-055. West arch, Eads Bridge, Mississippi River, St. Louis, Missouri, 1867–1874. James B. Eads, engineer-in-chief; James Andrews (Pittsburgh, PA), masonry contractor; Keystone Bridge Company (Pittsburgh, PA), prime contractor for iron and steel. Jet T. Lowe, photographer, 1983. P&P,HAER,MO,96-SALU,77-9.

Each of the three steel-arch spans is over 500 feet long.

2-056. Western approach viaduct, Eads Bridge, Mississippi River, St. Louis, Missouri, 1867–1874. James B. Eads, engineer-in-chief; James Andrews (Pittsburgh, PA), masonry contractor; Keystone Bridge Company (Pittsburgh, PA), prime contractor for iron and steel. Jet T. Lowe, photographer, 1983. P&P,HAER,MO,96-SALU, 77-17.

2-057. Arch connection at western pier, Eads Bridge, Mississippi River, St. Louis, Missouri, 1867–1874. James B. Eads, engineer-in-chief; James Andrews (Pittsburgh, PA), masonry contractor; Keystone Bridge Company (Pittsburgh, PA), prime contractor for iron and steel. Jet T. Lowe, photographer, 1983. P&P,HAER,MO,96-SALU,77-27.

Hollow wrought-iron tubes filled with chromium-steel staves compose the voussoirs of the arches. The other structural members are all wrought iron.

2-058. Railroad deck, Eads Bridge, Mississippi River, St. Louis, Missouri, 1867–1874. James B. Eads, engineer-in-chief; James Andrews (Pittsburgh, PA), masonry contractor; Keystone Bridge Company (Pittsburgh, PA), prime contractor for iron and steel. Jet T. Lowe, photographer, 1983. P&P,HAER,MO,96-SALU,77-35.

2-059

2-059. Washington Bridge, Harlem River, New York, New York, 1885–1889. C. C. Schneider, engineer; William J. McAlpine and William R. Hutton, chief engineers; Theodore Cooper, consulting engineer; Passaic Rolling Mill Company and Miles Tierney, contractors. William Henry Jackson, photographer, ca. 1890. P&P,LC-USZ62-93671.

The twin two-hinged steel arches of this highway bridge each span 510 feet. The Old Croton Aqueduct High Bridge (2-022) appears in the background.

2-060. Steel Arch Bridge, Mississippi River, Minneapolis, Minnesota, 1886–1891 (demolished 1987). Andrew Rinker, city engineer; Frederick Cappelen, city bridge engineer; north half designed by C. L. Strobell of Keystone Bridge Company; Keystone Bridge Company (Pittsburgh, PA) and Wrought Iron Bridge Company (Canton, OH), fabricators. Burt Levy, photographer, 1986. P&P,HAER,MINN,27-MINAP,11-6.

2-061. Upper Steel Arch Bridge, Niagara River, Niagara Falls, Niagara County, New York, 1897–1898 (collapsed 1938). Leffert L. Buck, chief engineer; Pencoyd Bridge Company (Philadelphia, PA), builder. Color photochrom print, DETR, ca. 1900. P&P, LOT 12006, p. 7.

The 840-foot-long, two-hinge, steel arch bridge replaced the suspension bridge, first built by John Roebling (see 5-030), which Buck had reconstructed in 1886 (see 5-032). Unusually heavy ice floes from Lake Erie in January 1938 damaged the steel structure bearing on the abutments, and the bridge collapsed. A new steel arch bridge, the Rainbow Bridge, was built to the north of the former structure, and it remains in use today.

2-060

2-061

2-062

2-062. Longfellow Bridge, Charles River, Boston, Massachusetts, 1900–1907. William Jackson, chief engineer; Edmund Wheelwright, architect. Jet T. Lowe, photographer, 1982. P&P,HAER,MASS,13-BOST,80-3.

2-063. Steel Bridge (Confederate Avenue Bridge), Vicksburg National Military Park, Vicksburg Vic., Warren County, Mississippi, 1903 (demolished ca. 2000). Penn Bridge Company (Beaver Falls, PA), fabricator. Jack E. Boucher, photographer, 1972. P&P,HAER,MISS,75-VICK,19-1.

2-064. Connection details, Steel Bridge (Confederate Avenue Bridge), Vicksburg National Military Park, Vicksburg Vic., Warren County, Mississippi, 1903 (demolished ca. 2000). Penn Bridge Company (Beaver Falls, PA), fabricator. Pete Brooks, delineator, 1997. P&P,HAER,MISS,75-VICK,19-,sheet no. 2.

This 150-foot deck-arch bridge has three pinned hinges located at the skewbacks and the apex of the arch. (A skewback is the end of an arch that inclines to meet the bearing and the abutment.)

2-063

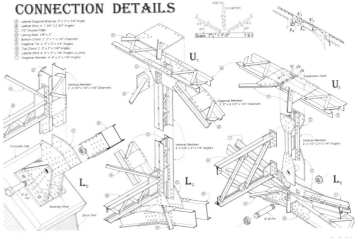

2-064

2-065

2-066

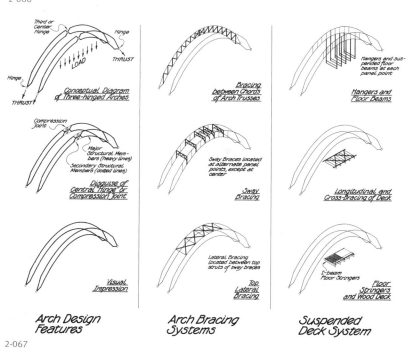

Arch Design Features
Arch Bracing Systems
Suspended Deck System

2-067

2-065. View from Bellows Falls, Bellows Falls Arch Bridge, Connecticut River, North Walpole, Cheshire County, New Hampshire, and Bellows Falls, Windham County, Vermont, 1905 (demolished 1982). J. R. Worcester, engineer; Louis A. Shoemaker & Company (Philadelphia, PA), steel fabricator. Jack Delano, photographer, 1941. P&P,LC-USF34-045254-D.

2-066. Bellows Falls Arch Bridge, Connecticut River, North Walpole, Cheshire County, New Hampshire, and Bellows Falls, Windham County, Vermont, 1905 (demolished 1982). J. R. Worcester, engineer; Louis A. Shoemaker & Company (Philadelphia, PA), steel fabricator. Jet T. Lowe, photographer, 1979.P&P,HAER,NH,3-WALPN, 1-11.

At the time of its completion, this 540-foot, three-hinged arch was the longest single-span highway bridge in the United States.

2-067. Structural subsystems, Bellows Falls Arch Bridge, Connecticut River, North Walpole, Cheshire County, New Hampshire, and Bellows Falls, Windham County, Vermont, 1905 (demolished 1982). J. R. Worcester, engineer; Louis A. Shoemaker & Company (Philadelphia, PA), steel fabricator. Richard K. Anderson Jr., delineator, 1982. P&P,HAER,NH,3-WALPN,1-,sheet no. 2.

Steel and concrete arch bridges often have hinges that allow the structure to compensate for stresses due to temperature changes, dynamic loads, or settlement in the abutments. The two-hinge arch, which locates the hinges at the skewbacks, is the most common configuration today; but in the early twentieth century, a third hinge often was added at the crown of the arch, as seen here.

2-068. Cedar Street Bridge, Kinnickinnic River, River Falls, Pierce County, Wisconsin, 1908. Minneapolis Steel & Machinery Company, builder. Martin Stupich, photographer, 1987. P&P,HAER, WIS,47-RIFA,1-3.

2-068

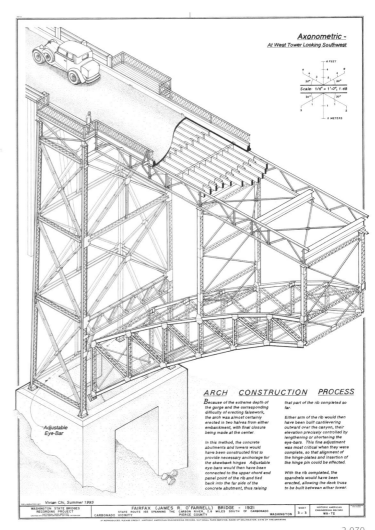

2-069

2-070

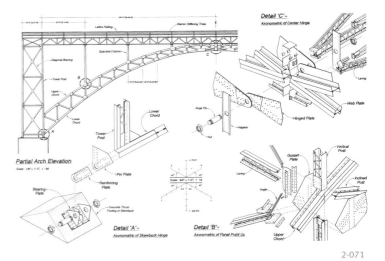

2-071

2-069. Fairfax Bridge (James R. O'Farrell Bridge), Carbon River, Carbonado Vic., Pierce County, Washington, 1921. E. A. White (Pierce County, WA), engineer, designer; Union Iron & Bridge Company (Seattle, WA), contractor; Minneapolis Steel & Machinery Company, steel fabricator. Jet T. Lowe, photographer, 1993. P&P,HAER,WASH,27-CARB.V,1-2.

2-070. Axonometric view of structure, Fairfax Bridge (James R. O'Farrell Bridge), Carbon River, Carbonado Vic., Pierce County, Washington, 1921. E. A. White (Pierce County, WA), engineer, designer; Union Iron & Bridge Company (Seattle, WA), contractor; Minneapolis Steel & Machinery Company, steel fabricator. Vivian Chi, delineator, 1993. P&P,HAER,WASH,27-CARB.V,1-,sheet no. 3.

2-071. Connection details showing three-hinge construction, Fairfax Bridge (James R. O'Farrell Bridge), Carbon River, Carbonado Vic., Pierce County, Washington, 1921. E. A. White (Pierce County, WA), engineer, designer; Union Iron & Bridge Company (Seattle, WA), contractor; Minneapolis Steel & Machinery Company, steel fabricator. Vivian Chi, delineator, 1993. P&P,HAER,WASH,27-CARB.V,1-,sheet no. 2.

2-072

2-073

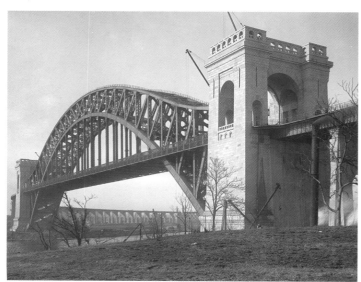

2-074

2-072. Aerial view, Hell Gate Bridge, New York Connecting Company, East River, New York, New York, 1904–1917. Gustav Lindenthal, chief engineer; Henry Hornbostel, consulting architect; American Bridge Company (Philadelphia, PA), builder, main span; McClintic-Marshall Company (Pittsburgh, PA), builder, approaches. Jet T. Lowe, photographer, 1991. P&P,HAER,NY,31-NEYO,176-3.

The bridge is the centerpiece of the rail network by which trains from New England and Long Island enter Manhattan. The full length of the four-tracked structure from Long Island to the Bronx is 17,000 feet; it includes a two-hinged arch spanning about 1,000 feet, the longest in the world at the time the bridge was completed.

2-073. Members of engineering staff, Hell Gate Bridge, New York Connecting Railroad Company, East River, New York, New York, 1904–1917. Gustav Lindenthal, chief engineer; Henry Hornbostel, consulting architect. Unidentified photographer, November 9, 1916. P&P,LC-USZ62-77070.

Lindenthal (1850–1935), regarded as the dean of American bridge engineering in the early twentieth century, stands at the center (white beard). To his right is his assistant, Othmar Ammann (1879–1965), who would achieve renown as the chief engineer for such other prominent New York City spans as the George Washington Bridge (1931) and the Verrazano–Narrows Bridge (1964).

2-074. Hell Gate Bridge, New York Connecting Railroad Company, East River, New York, New York, 1904–1917. Gustav Lindenthal, chief engineer; Henry Hornbostel, consulting architect; American Bridge Company (Philadelphia, PA), builder, main span; McClintic-Marshall Company (Pittsburgh, PA), builder, approaches. DETR, photographer, ca. 1915–1920. P&P,LC-D428-931.

2-075. Hell Gate Bridge during construction, New York Connecting Railroad Company, East River, New York, New York, 1904–1917. Gustav Lindenthal, chief engineer; Henry Hornbostel, consulting architect. Etching by Joseph Pennell, 1915. P&P,LC-USZ62-104967.

Lindenthal avoided the use of centering, which would have disrupted the shipping channel, by cantilevering the arch segments and supporting them with backstays (visible here at right) that were removed after the arch was closed.

2-076. Construction photograph, Hell Gate Bridge, New York Connecting Railroad Company, East River, New York, New York, 1904–1917. Gustav Lindenthal, chief engineer; Henry Hornbostel, consulting architect; American Bridge Company (Philadelphia, PA), builder, main span; McClintic-Marshall Company (Pittsburgh, PA), builder, approaches. Unidentified photographer, September 30, 1915. P&P,LC-USZ62-70740.

2-075

ARCH, HELL GATE BRIDGE 9/30/15

3646-1

2-076

2-077

2-077. Willamette River Bridge, Oregon City, Multnomah County, Oregon, 1922. Conde B. McCullough, state bridge engineer, Oregon State Highway Department; A. Guthrie & Company, Inc. (Portland, OR), builder. Jet T. Lowe, photographer, 1990. P&P,HAER,ORE, 3-ORGCI,2-4.

The 360-foot, three-hinged, steel box-girder through arch is encased in Gunite (a type of sprayed concrete) to protect the steel from the acidic fumes released by nearby paper mills. Built as part of a campaign upgrading the state highway system to accommodate rapidly increasing truck and automobile traffic, the bridge provided travelers with restrooms in both piers.

2-078

2-078. Crooked River Bridge, Terrebonne, Deschutes County, Oregon, 1926. Conde B. McCullough, state bridge engineer, Oregon State Highway Department; Kuckenberg & Wittman (Portland, OR), builder. Jet T. Lowe, photographer, 1990. P&P,HAER,ORE,9-TERBO.V,1-6.

2-079. Pedestrian overlook, Crooked River Bridge, Terrebonne, Deschutes County, Oregon, 1926. Conde B. McCullough, state bridge engineer, Oregon State Highway Department; Kuckenberg & Wittman (Portland, OR), builder. Jet T. Lowe, photographer, 1990. P&P,HAER,ORE,9-TERBO.V,1-13.

McCullough's design team selected materials for the overlook park that complemented the colors and textures of the site.

2-079

2-080. Navajo Bridge, Colorado River, Page Vic., Coconino County, Arizona, 1927–1929. Ralph A. Hoffman, Arizona Highway Department, bridge engineer; Kansas City Structural Steel Company, builder. Clayton B. Fraser, photographer, 1990. P&P,HAER,ARIZ,3-PAG.V,2-2.

2-081. French King Bridge, Connecticut River, Erving Vic., Franklin County, Massachusetts, 1932. George E. Harkness, bridge engineer, Massachusetts Department of Public Works; McClintic-Marshall Construction Company (Pittsburgh, PA), steelwork; Simpson Brothers Construction Company (Boston, MA), substructure. Martin Stupich, photographer, 1990. P&P,HAER,MASS,6-ERV,1-7.

2-082. Lorain-Carnegie Bridge, Cuyahoga River, Cleveland, Ohio, 1932. Wilbur J. Watson, consulting engineer; Frank Walker, consulting architect. Jet T. Lowe, photographer. P&P,HAER,OHIO,18-CLEV,39-2.

2-080

2-081

2-082

2-083

2-083. Yaquina Bay Bridge, Newport, Lincoln County, Oregon, 1934–1936. Conde B. McCullough, Oregon State Highway Department, state bridge engineer; Gilpin Construction Company (Portland, OR) and General Construction Company (Seattle, WA), builders. Jet T. Lowe, photographer, 1990. P&P,HAER,ORE,21-NEWPO,1-9.

The main 600-foot through arch is flanked by a pair of 350-foot steel deck arches. Five reinforced concrete deck arches complete the crossing of the Yaquina Bay on the Oregon Coastal Highway.

2-084. Yaquina Bay Bridge, Newport, Lincoln County, Oregon, 1934–1936. Conde B. McCullough, Oregon State Highway Department, state bridge engineer; Gilpin Construction Company (Portland, OR) and General Construction Company (Seattle, WA), builders. David Plowden, photographer, 1968. P&P,GFB-68-5518-K3,MphP732 B45.

Entry pylons are signature features of McCullough's major bridges.

2-084

2-085

2-086

2-085. Arch footing, Centennial Bridge, Mississippi River, Davenport, Scott County, Iowa, 1939–1940. Edward L. Ashton, engineer; American Bridge Company (Pittsburgh, PA), superstructure; McCarthy Improvement Company (Davenport, IA), substructure; Crouse & Sanders (Detroit, MI), deck. Joseph Elliott, photographer, 1995. P&P,HAER,IOWA,82-DAVPO,8-9.

Tie beams beneath the deck (at left) of each of the five spans of this 2,262-foot bridge resist the lateral forces of the arches and allow for much thinner piers.

2-086. Centennial Bridge, Mississippi River, Davenport, Scott County, Iowa, 1939–1940. Edward L. Ashton, engineer; American Bridge Company (Pittsburgh, PA), superstructure; McCarthy Improvement Company (Davenport, IA), substructure; Crouse & Sanders (Detroit, MI), deck. Joseph Elliott, photographer, 1995. P&P,HAER,IOWA,82-DAVPO,8-4.

2-087. Toll booth, Centennial Bridge, Mississippi River, Davenport, Scott County, Iowa, 1939–1940. Edward L. Ashton, engineer; American Bridge Company (Pittsburgh, PA), superstructure; McCarthy Improvement Company (Davenport, IA), substructure; Crouse & Sanders (Detroit, MI), deck. Joseph Elliott, photographer, 1995. P&P,HAER,IOWA,82-DAVPO,8-7.

2-088. McKee's Rocks Bridge, Ohio River, McKee's Rocks and Pittsburgh, Allegheny County, Pennsylvania, 1929–1931. V. R. Covell, county engineer; Allegheny County Department of Public Works, designer; Fort Pitt Bridge Works (Pittsburgh, PA), superstructure; Dravo Engineering Works (Pittsburgh, PA), substructure. Joseph Elliott, photographer, 1997. P&P,HAER,PA,2-MCKRO,2-3.

2-087

2-088

2-089

2-091

2-090

2-089. Bayonne Bridge, Kill Van Kull, Bayonne, Hudson County, New Jersey, 1927–1931. Othmar Ammann, chief engineer; Cass Gilbert, consulting architect; American Bridge Company (Pittsburgh, PA) builder. Margaret Bourke-White, photographer, ca. 1940. P&P,LC-USZ62-83230.

2-090. Aerial view, Bayonne Bridge, Kill Van Kull, Bayonne, Hudson County, New Jersey, 1927–1931. Othmar Ammann, chief engineer; Cass Gilbert, consulting architect; American Bridge Company (Pittsburgh, PA), builder. Unidentified photographer. P&P,HAER,NJ,9-BAYO,1-18.

With a clear span of 1,675 feet, this was the world's longest arch bridge until the New River Gorge Bridge (2-093–2-094) surpassed it in 1977.

2-091. Walkway, suspender cables, and deck beams, Bayonne Bridge, Kill Van Kull, Bayonne, Hudson County, New Jersey, 1927–1931. Othmar Ammann, chief engineer; Cass Gilbert, consulting architect; American Bridge Company (Pittsburgh, PA), builder. Jet T. Lowe, photographer, 1985. P&P,HAER,NJ,9-BAYO,1-6.

2-092

2-093

2-092. Notre Dame Bridge, Merrimack River, Manchester, Hillsborough County, New Hampshire, 1936–1937 (demolished 1988). J. R. Worceser Company (Boston, MA), engineers; American Bridge Company (Pittsburgh, PA), superstructure. Bruce Alexander and Ernest Gould, photographers, 1988. P&P,HAER,NH,6-MANCH,12-3.

2-093. Aerial view, New River Gorge Bridge, Fayetteville Vic., Fayette County, West Virginia, 1973–1977. Michael Baker Company, engineers; American Bridge Division, U.S. Steel (Pittsburgh, PA), builder. Jet T. Lowe, photographer, 1988. P&P,HAER,WVA,10-FAY.V,1-10.

The 1,700-foot deck arch of Cor-ten steel spans the gorge 876 feet above the New River. Cor-ten is a steel alloy produced by U.S. Steel that oxidizes to form a corrosion-resistant surface.

2-094. New River Gorge Bridge, Fayetteville Vic., Fayette County, West Virginia, 1973–1977. Michael Baker Company, engineers; American Bridge Division, U.S. Steel (Pittsburgh, PA), builder. Jet T. Lowe, photographer, 1988. P&P,HAER,WVA,10-FAY.V,1-2.

2-094

CONCRETE ARCH BRIDGES

The use of concrete as a structural material in the United States increased dramatically during the 1870s. In bridge construction, the first concrete arch was built in New York's Prospect Park in 1871, and the material was used occasionally in combination with stone masonry (2-025, 2-043). Its application became increasingly widespread with the introduction of techniques for reinforcing it with steel in the 1880s.

2-095

2-095. Alvord Lake Bridge, Golden Gate Park, San Francisco, California, 1889. Ernest L. Ransome, engineer. Jet T. Lowe, photographer, 1984. P&P,HAER, CAL,38-SANFRA,138-3.

This 20-foot span was the first reinforced concrete arch to be built in the United States. Ransome (1852–1917), a pioneer of concrete construction, reinforced the concrete with twisted steel rods, the forerunners of today's rebar.

2-096. Alvord Lake Bridge, Golden Gate Park, San Francisco, California, 1889. Ernest L. Ransome, engineer. Jet T. Lowe, photographer, 1984. P&P,HAER, CAL,38-SANFRA,138-4.

The facades of the bridge are textured to recall stone masonry. Decorative concrete stalactites hang from the ceiling.

2-097. W. H. Pratt Bridge (Gaddis Ford Bridge), tributary of North Branch Kokosing River, Fredericktown Vic., Knox County, Ohio, 1896. Buckeye Portland Cement Company (Bellefontaine, OH), builder. Joseph Elliott, photographer, 1992. P&P,HAER,OHIO,42-FRED.V,1-2.

2-096

2-097

2-098. Cooper Street Bridge, spanning Massachusetts Bay Transportation Authority tracks (former Boston & Maine Railroad), Wakefield, Middlesex County, Massachusetts, 1903. Joseph Ross, builder, 1903. Wayne Fleming, photographer, 1995. P&P,HAER, MASS,9-WAK,2-1.

2-099. Union Pacific Railroad Bridge (Los Angeles & Salt Lake Railroad Bridge), Santa Ana River, Riverside Vic., Riverside County, California, 1902–1904. E. B. & A. L. Stone Company (Oakland, CA), builders. Brian Grogan, photographer, 1991. P&P,HAER, CAL,33-RIVSI.V,1-1.

This 984-foot viaduct composed of eight 86-foot spans and two 38-foot spans of unreinforced concrete illustrates the growing confidence in concrete construction at the turn of the twentieth century.

2-100. Morton's Highway Crossing (O Avenue Bridge), spanning Amtrak tracks, Oshtemo Vic., Kalamazoo County, Michigan, 1904. B. Douglas, Michigan Central Railroad, bridge engineer. Dietrich Floeter, photographer, 1990. P&P,HAER,MICH,39-OSHT.V,1-6.

2-098

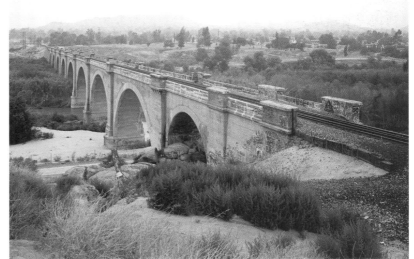

2-099

2-100

2-101

2-101. Rocky River Bridge, Rocky River, Cuyahoga County, Ohio, 1908–1910. A. M. Felgate, county bridge engineer, designer; Schillinger Brothers Company (Columbus, OH), builder. DETR, photographer, ca. 1910–1920. P&P,LC-D4-72379.

At the time of its completion, the 280-foot span of the Rocky River Bridge was the longest concrete arch in the world. It is one of about a dozen surviving mass-poured (unreinforced) concrete bridges in America (see 2-099).

2-102. View of main arch ribs, Rocky River Bridge, Rocky River, Cuyahoga County, Ohio, 1908–1910. A. M. Felgate, county bridge engineer, designer; Schillinger Brothers Company (Columbus, OH), builder. Mike Fuerst, photographer, 1976. P&P,HAER,OHIO,18-RORI,1-20.

Felgate adapted a German system of unreinforced concrete construction that offered savings of materials and weight by reducing the primary structure to two relatively narrow parallel arches.

2-103. Galveston Railroad Causeway, Galveston Bay, Galveston, Galveston County, Texas, 1909–1912. Concrete Steel Engineering Company (New York, NY), designer; A. M. Blodgett Construction Company (Kansas City, MO), builder. Joseph Elliott, photographer, 1996. P&P,HAER,TEX,84-GALV,43-1.

Built as part of the reconstruction following the hurricane of 1900, the 2-mile reinforced concrete causeway links Galveston Island to the mainland.

2-104. Benson Street Bridge, Mill Creek, Lockland, Hamilton County, Ohio, 1909–1910. E. A. Gast, deputy county surveyor, designer; Peter Praechter, builder. Louise T. Cawood, photographer, 1986. P&P,HAER,OHIO,31-LOCK,1-3.

This is an early example of a concrete bowstring arch, a structural form that would be widely used in the 1920s by engineers such as James Marsh (2-153–2-158, 2-160–2-163).

2-102

2-103

2-104

2-105. Delaware River Bridge (Yardley Bridge), Philadelphia & Reading Railroad, Yardley, Bucks County, Pennsylvania, 1911–1913. William Hunter, chief engineer, and Edwin Chamberlin, assistant engineer; Philadelphia & Reading Railroad, designers; F. W. Talbot Construction Company (New York, NY), builder. Joseph Elliott, photographer, 1999. P&P,HAER,PA,9-YARD,8-1.

2-106. Gull Street Bridge, Kalamazoo River, Kalamazoo, Kalamazoo County, Michigan, 1911. H. A. Johnson, city engineer, designer; Richard Heystek Sr., builder. Howard Wright, photographer, 1992. P&P,HAER,MICH,39-KALAM,5-4.

The bridge has three 59-foot arches composed of reinforced concrete shells filled with packed earth that provides the base for the road surface.

2-107. Colorado Street Bridge, Arroyo Seco, Pasadena, Los Angeles County, California, 1912–1913 (rehabilitated 1988–1990). J. A. L. Waddell, engineer; John Drake Mercerau, builder. Tavo Olmos, photographer, 1988. P&P,HAER,CAL,19-PASA,11-8.

Built as part of the county boulevard system in response to growing automobile traffic, the bridge has open spandrels, which reduce its weight. The classical moldings on the spandrel columns complement the ornamental railings and street lamps on the deck.

2-105

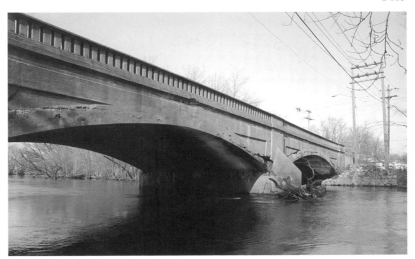

2-106

2-107

2-108. Oak Cliff Viaduct, Trinity River, Dallas, Texas, 1912. Ira G. Hedrick, engineer. Johnson & Rogers, photographer, 1912. P&P, PAN US GEOG Texas, no. 73 (E size).

This is an early independent work of Ira G. Hedrick, best known for the large bridges he designed in Arkansas in the 1920s and 1930s (see 3-375–3-378).

2-109. Parks Bar Bridge, Yuba River, Smartville Vic., Yuba County, California, 1913 (demolished 1993). William M. Thomas, engineer; Portland Concrete Pile Company, builder. Dan Tateishi, photographer, 1991. P&P,HAER,CAL,58-SMAVI.V,1-9.

The bridge is an example of the Thomas System, patented in 1908, which employed three-hinged, pre-cast, reinforced concrete arches. William Thomas and his partner, W. S. Post, built at least eighteen concrete bridges in California.

2-110. Presumpscot Falls Bridge, Presumpscot River, Falmouth, Cumberland County, Maine, 1913. Sanders Contracting Company, designer and builder. Brian Vandenbrink, photographer, 1993. P&P,HAER,ME,3-FAL,1-7.

2-111. Shepperds Dell Bridge, Young Creek, Latourell, Multnomah County, Oregon, 1914. K. P. Billner, bridge engineer; Samuel C. Lancaster, consulting engineer; Pacific Bridge Company (Portland, OR), builder. Jet T. Lowe, photographer, 1990. P&P,HAER,ORE,26-LATO.V,1-1.

The Shepperds Dell Bridge's main span is a 100-foot, reinforced concrete, ribbed deck arch. It is one of the dozen structures Billner designed for the Columbia River Highway, America's first scenic highway.

2-112. Fromberg Bridge, Clark's Fork of Yellowstone River, Fromberg Vic., Carbon County, Montana, 1914. C. A. Gibson, county surveyor, designer; Beley Construction Company (Livingston, MT), builder. Jet T. Lowe, photographer, 1980. P&P,HAER,MONT,5-FROBE.V,1-5.

2-113. Tunkhannock Viaduct, Erie Lackawanna Railroad, Tunkhannock Creek valley, Nicholson, Wyoming County, Pennsylvania, 1915. A. Burton Cohen, design engineer; George J. Ray, chief engineer; David W. Flickmir and Lincoln Bush, builders. Jet T. Lowe, photographer, 1993. P&P,HAER,PA,66-NICH,1-7.

The 2,375-foot structure spans the valley with ten 180-foot reinforced concrete, open spandrel arches.

2-111

2-112

2-113

2-114

2-115

2-116

2-114. Horner Street Bridge, Johnstown, Cambria County, Pennsylvania, 1916–1917. Gustav A. Fink, engineer; John L. Elder, builder. Robert C. Shelley, photographer, 1991. P&P,HAER,PA,11-JOTO,138-11.

2-115. Court Street Bridge, Black River, Watertown, Jefferson County, New York, 1922 (demolished 1989). William Mueser and Mark D. Ewell, Concrete-Steel Engineering Company (New York, NY), designers; Peckham Construction Company (Buffalo, NY), builder. Martin Stupich, photographer, 1987. P&P,HAER,NY,23-WATO,5-2.

2-116. Hayden Arch Bridge, Shoshone River, Cody Vic., Park County, Wyoming, 1924–1925. C. E. Hayden and J. F. Seiler, Wyoming State Highway Department, engineers; H. S. Crocker (Denver, CO), builder. Clayton B. Fraser, photographer, 1982. P&P,HAER,WYO,15-CODY.V,1-2.

2-117. Mosel Avenue Bridge, Kalamazoo River, Kalamazoo, Kalamazoo County, Michigan, 1924 (demolished 1990). Michigan State Highway Department, designer; J. P. Rusche (Grand Rapids, MI), builder. Dietrich Floeter, photographer, 1989. P&P,HAER, MICH,39-KALAM,2-1.

2-118. Ninth Street Viaduct, Los Angeles River, Los Angeles, California, 1925. Merrill Butler, chief engineer, City of Los Angeles, designer; North Pacific Construction Company, builder. Tavo Olmos, photographer, 1996. P&P,HAER,CAL,19-LOSAN,78-6.

2-119. Market Street Bridge, Susquehanna River, Harrisburg, Pennsylvania, 1926–1928. Ralph Modjeski and Frank N. Masters, engineers; Paul Cret, architect; James McGraw Company, contractor. Joseph Elliott, photographer, 1997. P&P,HAER, PA,22-HARBU,27-1.

2-117

2-118

2-119

2-120

2-120. Soldiers' and Sailors' Memorial Bridge, Paxton Creek, Harrisburg, Pennsylvania, 1926–1930. John E. Greiner, consulting engineer; William Gehron and Sidney F. Ross, architects; Charles C. Strayer and James McGraw Company, contractors. Joseph Elliott, photographer, 1997. P&P,HAER,PA,22-HARBU,28-1.

The bridge was designed as part of the approach to the state capitol. In keeping with its monumental setting, the reinforced concrete structure is faced with limestone ashlar. The pylons (the towers flanking the approach) and the military imagery carved on the keystones (voussoirs at the top of arches) of the arches denote the bridge's dedication to the state's veterans. A military museum incorporated into the structure was planned but not realized.

2-121. Georgia-Carolina Memorial Bridge, Savannah River, Elberton Vic., Elbert County, Georgia, 1926–1927. Searcy B. Slack, engineer; Emmett M. Williams, contractor. Dennis O'Kain, photographer, 1980. P&P,HAER,GA,53-ELBE.V,2-7.

2-122. Gervais Street Bridge, Congaree River, Columbia, Richland County, South Carolina, 1926–1927. Joseph W. Barswell, engineer. Jack E. Boucher, photographer, 1986. P&P,HAER,SC,40-COLUM,19-8.

2-123. Street lamp, Gervais Street Bridge, Congaree River, Columbia, Richland County, South Carolina, 1926–1927. Joseph W. Barswell, engineer. Jack E. Boucher, photographer, 1986. P&P,HAER, SC,40-COLUM,19-10.

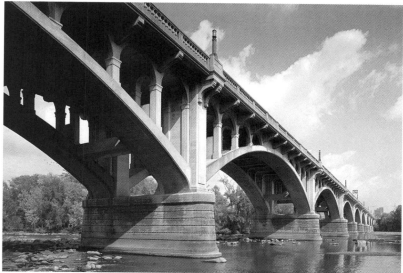
2-121

2-122

2-123

2-124. George Westinghouse Bridge, Turtle Creek, East Pittsburgh, Allegheny County, Pennsylvania, 1929–1932. Vernon R. Covell, George H. Richardson, A. D. Nutter, Allegheny County Department of Public Works, engineers; Booth & Flinn Company (Pittsburgh, PA), contractor. Joseph Elliott, photographer, 1997. P&P,HAER,PA,2-EAPIT,1-3.

Five reinforced concrete deck arches (the longest span is 460 feet) compose this 1,560-foot bridge carrying the Lincoln Highway above Turtle Creek and railroad tracks serving the Westinghouse Electric Company.

2-125. View of deck, George Westinghouse Bridge, Turtle Creek, East Pittsburgh, Allegheny County, Pennsylvania, 1929–1932. Vernon R. Covell, George H. Richardson, A. D. Nutter, Allegheny County Department of Public Works, engineers; Booth & Flinn Company (Pittsburgh, PA), contractor. Joseph Elliott, photographer, 1997. P&P,HAER,PA,2-EAPIT,1-7.

2-126. Pylon with sculpture, George Westinghouse Bridge, Turtle Creek, East Pittsburgh, Allegheny County, Pennsylvania, 1933–1936. Stanley Roush, architect; Frank Vittor, sculptor. Joseph Elliott, photographer, 1997. P&P,HAER,PA,2-EAPIT,1-9.

The subjects of Vittor's bas-reliefs include transportation, electricity, steel, and the settlement of Turtle Creek Valley.

2-125

2-126

2-127

2-127. Arlington Memorial Bridge, Potomac River, Washington, D.C., 1929–1932. McKim, Mead & White, architects, William Mitchell Kendell, project architect; John L. Nagle, engineer; W. J. Douglas, consulting engineer; Joseph P. Strauss, bascule span engineer. Jack E. Boucher, photographer, 1992. P&P,HAER,DC,WASH,563-20.

Designed to harmonize with the Lincoln Memorial and the classical aesthetic mandated by the National Commission of Fine Arts, the Arlington Memorial Bridge consists of eight concrete arches faced with granite and a double-leaf steel bascule span faced with pressed steel.

2-128

2-128. Centering for concrete arch, Arlington Memorial Bridge, Potomac River, Washington, D.C., 1929–1932. McKim, Mead & White, architects, William Mitchell Kendell, project architect; John L. Nagle, engineer. Theodor Horydczak, photographer, ca. 1930. P&P,LC-H824-3637-x.

2-129. Setting voussoir veneer stone to concrete core, Arlington Memorial Bridge, Potomac River, Washington, D.C., 1929–1932. McKim, Mead & White, architects, William Mitchell Kendell, project architect; John L. Nagle, engineer. Theodor Horydczak, photographer, ca. 1930. P&P,LC-H824-T-3521-x.

2-129

2-130

2-130. Construction of bascule span, Arlington Memorial Bridge, Potomac River, Washington, D.C., 1929–1932. McKim, Mead & White, architects, William Mitchell Kendell, project architect; John L. Nagle, engineer; Joseph P. Strauss, bascule span engineer. Theodor Horydczak, photographer, ca. 1930. P&P,LC-H814-T-3643-x.

2-131. Roadway construction, Arlington Memorial Bridge, Potomac River, Washington, D.C., 1929–1932. McKim, Mead & White, architects, William Mitchell Kendell, project architect; John L. Nagle, engineer. Theodor Horydczak, photographer, ca. 1930. P&P,LC-H824-T-3525-x.

2-132. Assembly of granite balustrade, Arlington Memorial Bridge, Potomac River, Washington, D.C., 1929–1932. McKim, Mead & White, architects, William Mitchell Kendell, project architect; John L. Nagle, engineer. Theodor Horydczak, photographer, ca. 1930. P&P, LC-H824-T-3529-x.

2-131

2-132

2-133

2-134

2-133. Narrows Bridge, Juniata River, Bedford Vic., Bedford County, Pennsylvania, 1934–1935. E. E. Brandon, Pennsylvania Division of Highways, engineer; Pittsburgh Construction Company, contractor. Joseph Elliott, photographer, 1997. P&P,HAER,PA,5-BED.V,1-5.

The site and approach of the Lincoln Highway (Route 30) required the five-span (each 119 feet) open-spandrel bridge to be built at a skew with a curve and a rising slope.

2-134. Hunting Creek Bridge, Mount Vernon Memorial Highway, Alexandria Vic., Fairfax County, Virginia, 1929–1932. E. T. Larsen, E. W. Armstrong, J. V. McNary, U.S. Bureau of Public Roads, engineers; Gilmore D. Clarke, architect; Merritt-Chapman & Scott Corporation (New York, NY), contractor. Jet T. Lowe, photographer, 1989. P&P,HAER,VA,30-__,6-A-2.

The reinforced concrete structure is faced with local stone in keeping with the "rustic style" commonly employed by the Bureau of Public Roads and the National Park Service during the 1930s.

2-135. Spout Run Arch Bridge, Spout Run, eastbound span, George Washington Memorial Parkway, Arlington Vic., Arlington County, Virginia, 1959. R. H. Wood and H. J. Spelman, U.S. Bureau of Public Roads, engineers; William Hausmann, National Park Service, architect; Capital Engineering Company (Washington, DC), contractor. Jet T. Lowe, photographer, 1994. P&P,HAER,VA,7-ARL.V,14-1.

In contrast to the "rustic style," according to which earlier parkway bridges in the Washington area were clad with a veneer of local stone, the designers of this 222-foot, open-spandrel deck arch exposed the reinforced concrete structure as an expression of visual lightness and modernity.

2-136. Natchez Trace Parkway Bridge, Franklin Vic., Williamson County, Tennessee, 1994. Figg Engineering Group, designer; PCL Civil Constructors (Coral Gables, FL), builder. Figg Engineering Group, photographer, 1994. P&P,LC-DIG-ppem-00054.

A project of the Eastern Federal Lands Division of the Federal Highway Administration, this bridge was one of the final components of the 444-mile parkway (1938–2005) running between Natchez, Mississippi, and Nashville, Tennessee. Its two pre-cast segmental concrete arches, the first to be built in the United States, have spans of 582 and 462 feet.

2-137. Construction photo, Natchez Trace Parkway Bridge, Franklin Vic., Williamson County, Tennessee, 1994. Figg Engineering Group, designer; PCL Civil Constructors (Coral Gables, FL), builder. Figg Engineering Group, photographer, 1994. P&P,LC-DIG-ppem-00055.

2-135

2-136

2-137

The Melan Reinforcing System

In 1892, the Austrian engineer Josef Melan (1853– 1941) invented a system of concrete construction reinforced by steel beams, for which he subsequently obtained a patent in the United States. His first license holder was another Austrian engineer, Fritz von Emperger (1862–1942), who had been working in America since 1890. Von Emperger and others promoted the Melan system, and its use became widespread in the first years of the 20th century.

2-138

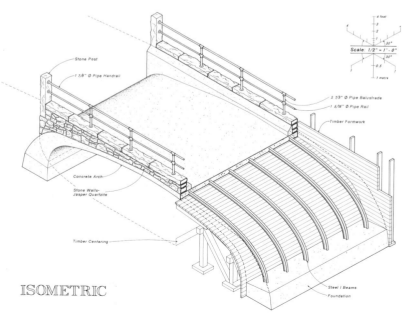

ISOMETRIC

2-139

2-138. Melan Arch Bridge, Emma Slater Park (moved from Dry Run Creek in 1964), Rock Rapids Vic., Lyon County, Iowa, 1893. Fritz von Emperger, engineer; W. S. Hewett (Minneapolis, MN), builder; John Olsen (Rock Rapids, IA), builder, 1893. Joseph Elliott, photographer, 1995. P&P,HAER,IOWA,60-ROCRA.V,1-3.

2-139. Isometric view, Melan Arch Bridge, Emma Slater Park (moved from Dry Run Creek in 1964), Rock Rapids Vic., Lyon County, Iowa, 1893. Fritz von Emperger, engineer; W. S. Hewett (Minneapolis, MN), builder; John Olsen (Rock Rapids, IA), builder, 1893. Erick McEvoy, delineator, 1995. P&P, HAER,IOWA,60-ROCRA.V,1-,sheet no. 2.

2-140. Boulder Bridge, Rock Creek, Rock Creek National Park, Washington, D.C., 1901–1902. Lansing H. Beach, engineer, designer; John Watkinson Douglas, District Bridge Commissioner, engineer; Taity & Allen (Washington, DC), contractor. Jet T. Lowe, photographer. P&P,HAER,DC,WASH,564-1.

2-141. Cutaway view, Boulder Bridge, Rock Creek, Rock Creek National Park, Washington, D.C., 1901–1902. Lansing H. Beach, engineer, designer; John Watkinson Douglas, District Bridge Commissioner, engineer; Taity & Allen (Washington, DC), contractor. Elaine Pierce, delineator, 1995. P&P,HAER,DC,WASH,564-,sheet no. 2.

Unlike the Rock Rapids bridge (2-138–2-139), which uses bent I-beams, the reinforcement here consists of steel trusses that follow the curve of the arches. The outer trusses support a veneer of local boulders.

2-142. Frankford Avenue Bridge, Poquessing Creek, Philadelphia, Pennsylvania, 1904. George S. Webster, Henry H. Quimby, engineers; John McMenamy, contractor. Joseph Elliott, photographer, 1997. P&P,HAER,PA,51-PHILA,705-2.

The exterior finish of the concrete is scored to give the appearance of stone masonry.

2-140

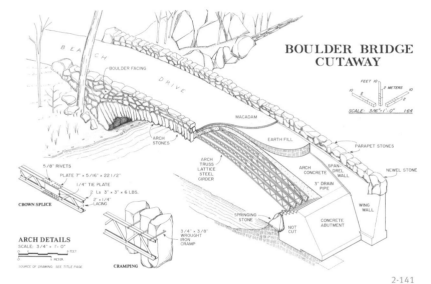

2-141

2-142

2-143

2-143. Sandy Hill Bridge, Hudson River, Hudson Falls, Washington County, New York, 1906–1997. M. O. Kasson, Union Bag & Paper Company, chief engineer; William Burr, Columbia University, consulting engineer; Union Bag & Paper Company, builder. Martin Stupich, photographer, 1987. P&P,HAER,NY,58-HUFA,1-4.

2-144. Wells Street Bridge, Spy Run Creek, Fort Wayne, Allen County, Indiana, 1914. O. B. Wiley, engineer; Burk Construction Company (New Castle, IN), contractor. Alan Conant, photographer, 1994. P&P, HAER,IND,2-FOWA,5-7.

2-145. Hampden County Memorial Bridge, Connecticut River, Springfield, Hampden County, Massachusetts, 1922. Fay, Spofford & Thorndike (Boston, MA), engineers; Haven & Hoyt (Boston, MA), consulting architects; H. P. Converse & Company, contractor. Martin Stupich, photographer, 1990. P&P,HAER,MASS,7-SPRIF,7-4.

2-144

2-145

Luten Patent Bridges

Daniel Luten (1869–1946) was an engineer and entrepreneur who at the turn of the twentieth century understood the potential commercial gains to be had by controlling the patents governing the design and construction processes of reinforced concrete bridges. By the 1920s, he had acquired over fifty patents and issued licenses for or directly super-vised the construction of more than 14,000 bridges through a series of corporations including the National Bridge Company and the Luten Bridge Company. Luten patents covered so many facets of bridge construction that other builders fre-quently found themselves facing lawsuits by his attorneys, who aggressively sought to identify potential infringements. Countersuits brought by competitors and state highway departments seeking to standardize bridge design in their jurisdictions eventually weakened Luten's dominance.

2-146

2-147

2-146. Commercial Street Bridge, Purgatoire River, Trinidad, Las Animas County, Colorado, 1905. Daniel Luten, design patent; Marsh Bridge Company, builder. Vic Macaluse, photographer, 1988. P&P,HAER,COLO,36-TRIN,3-2.

Luten favored elliptical arches as economical and aesthetically satisfying. His steel reinforcement was a variant of the Melan system. The builder of this bridge, James Marsh (2-157–2-167), became a major competitor of Luten in the 1910s.

2-147. Twin Bridge, Volga River, Fayette Vic., Fayette County, Iowa, ca. 1910. Daniel B. Luten, designer; N. M. Stark & Company (Des Moines, IA), contractor. Joseph Elliott, photographer, 1995. P&P,HAER,IOWA,33-FAY.V,1-1.

2-148. Isometric drawing showing construc-tion sequence, Twin Bridge, Volga River, Fayette Vic., Fayette County, Iowa, ca. 1910. Daniel B. Luten, designer; N. M. Stark & Company (Des Moines, IA), contractor. Min Xu, delineator, 1996. P&P,HAER,IOWA,33-FAY.V,1-,sheet no. 3.

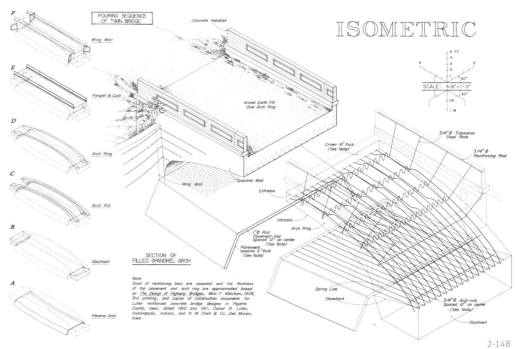

2-148

2-149

2-149. Putnam County Bridge No. 111, Little Walnut Creek, Greencastle Vic., Putnam County, Indiana, 1909. Daniel Luten, designer; S. H. Gibson, builder. Camille B. Fife, photographer, 1998. P&P,HAER,IND,67-GREC.V,1-8.

2-150. Luten's Bridge, Pine Log Creek, Cash Vic., Gordon County, Georgia, 1920. Daniel Luten, Luten Bridge Company, designer and builder. Brad Sprinkle, photographer, 1985. P&P,HAER,GA,65-CASH.V,1-1.

2-151. Illinois River Bridge (Midway Bridge), Siloan Springs Vic., Benton County, Arkansas, 1922. Daniel Luten, Luten Bridge Co., designer and builder. Louise T. Cawood, photographer, 1988. P&P,HAER,ARK,4-SISP,1-5.

2-152. Andrew J. Sullivan Memorial Bridge, Cumberland River, Williamsburg Vic., Whitley County, Kentucky, 1928. Daniel Luten, Luten Bridge Company, designer and builder. Jayne Fiegel and Kurt Fiegel, photographers, ca. 1999. P&P,HAER,KY,118-WILBU.V,2-1.

2-150

2-151

2-152

Arch Bridges by James Barney Marsh

James Barney Marsh (1856–1936) was an engineer based in Des Moines, Iowa, who began his career designing and building metal truss bridges. In the first years of the twentieth century, he became a proponent of reinforced concrete bridges, building them to his own designs and under license from Daniel Luten. In 1912, he obtained his first patent for the design of the "rainbow" arch bridges (so-named for their distinctive profiles) for which he became famous, building hundreds of the structures primarily in the Midwest.

2-153. Lake City Bridge, North Raccoon River, Lake City Vic., Calhoun County, Iowa, 1914. James B. Marsh, designer; Iowa Bridge Company (Des Moines, IA), builder. Joseph Elliott, photographer, 1995. P&P,HAER,IOWA,13-LACIT.V,1-3.

2-154. Lake City Bridge, North Raccoon River, Lake City Vic., Calhoun County, Iowa, 1914. James B. Marsh, designer; Iowa Bridge Company (Des Moines, IA), builder. Joseph Elliott, photographer, 1995. P&P,HAER,IOWA,13-LACIT.V,1-4.

2-155. Cutaway view, Lake City Bridge, North Raccoon River, Lake City Vic., Calhoun County, Iowa, 1914. James B. Marsh, designer; Iowa Bridge Company (Des Moines, IA), builder. Adriaan Vlaardingerbroek, delineator, 1995. P&P,HAER,IOWA,13-LACIT.V,1-,sheet no. 2.

This is an excellent example of Marsh's "rainbow" design, by which the deck is suspended from a heavy steel frame encased in concrete, a variation of the Melan system of reinforcement.

2-153

2-154

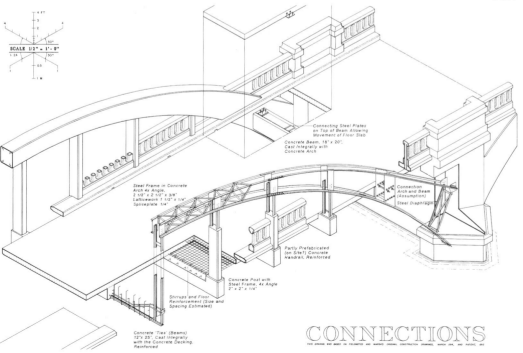

CONNECTIONS

2-155

2-156

2-157

2-158

2-156. View of deck, Spring Street Bridge, Duncan Creek, Chippewa Falls, Chippewa County, Wisconsin, 1916. James B. Marsh, designer; Iowa Bridge Company, builder (Des Moines, IA). Martin Stupich, photographer, 1987. P&P,HAER,WIS,9-CHIFA,2-1.

2-157. Oblique view, Spring Street Bridge, Duncan Creek, Chippewa Falls, Chippewa County, Wisconsin, 1916. James B. Marsh, designer; Iowa Bridge Company, builder (Des Moines, IA). Martin Stupich, photographer, 1987. P&P,HAER,WIS,9-CHIFA,2-2.

2-158. Court Avenue Bridge, Des Moines River, Des Moines, Iowa, 1918. Marsh Engineering Company, designer; Koss Construction, contractor. Joseph Elliott, photographer, 1995. P&P,HAER,IOWA,77-DESMO,25-5.

In keeping with its prominent site in the city, this reinforced concrete, elliptical-arch bridge has a classically detailed limestone auhlar facing.

2-159. Mederville Bridge, Volga River, Mederville, Clayton County, Iowa, 1918. Marsh Engineering Company, designer; F. E. Marsh & Company, builder. Bruce A. Harms, photographer, 1996. P&P,HAER,IOWA,22-MEDE,1-4.

Marsh employed the Melan system for the reinforcement of this 155-foot, open-spandrel deck arch bridge.

2-160. Mott Rainbow Arch Bridge, Cannonball River, Mott, Hettinger County, North Dakota, 1921 (demolished 1980). Marsh Engineering Company, designer; N. M. Stark & Company, builder. Garry Redmann, photographer, 1980. P&P,HAER,ND,21-MOTT, 1-11.

2-161. Bladensburg Concrete Bowstring Bridge, Wakatomika Creek, Bladensburg Vic., Knox County, Ohio, 1928. J. B. Marsh patent, Ohio Bureau of Bridges, designer; L. D. Kear (Wharton, OH), builder. Louise T. Cawood, photographer, 1986. P&P,HAER, OHIO,42-BLAD.V,1-3.

In 1921, the Ohio Department of Highways adopted Marsh's "rainbow" design as one of its standard bridge types.

2-159

2-160

2-161

2-162. Aerial view, Cotter Bridge (R. M. Ruthven Bridge), White River, Cotter, Baxter County, Arkansas, 1930. Marsh Engineering Company, designer; Bateman Contracting (Nashville, TN), builder. Louise Taft, photographer, 1988. P&P,HAER,ARK,3-COT,1-4.

2-163. Cotter Bridge (R. M. Ruthven Bridge), White River, Cotter, Baxter County, Arkansas, 1930. Marsh Engineering Company, designer; Bateman Contracting (Nashville, TN), builder. Louise Taft, photographer, 1988. P&P,HAER,ARK,3-COT,1-5.

Concrete Arch Bridges by Conde B. McCullough

Conde B. McCullough (1887–1946) worked briefly for James B. Marsh and then for the Iowa State Highway Commission before moving to Oregon to teach structural engineering at Oregon Agricultural College (now Oregon State University). In 1919, the Oregon State Highway Commission hired him as its bridge designer, and he stayed with the agency for the remainder of his career, having responsibility for the construction of nearly six hundred bridges as part of the state's campaign to create one of the nation's leading highway systems. McCullough designed a number of steel bridges (2-077–2-079, 2-083–2-084) but favored reinforced concrete arches as the structural system most amenable to designs that would be economical, structurally elegant, and aesthetically pleasing.

2-164. Rock Point Arch Bridge, Rogue River, Jackson County, Oregon, 1919–1920. Conde B. McCullough, designer; Parker & Banfield (Portland, OR), builder. Jet T. Lowe, photographer, 1990. P&P,HAER,ORE,15-GOLHI,2-3.

This was McCullough's first major bridge as designer for the Oregon State Highway Commission. The main span is a 113-foot, reinforced concrete, open-spandrel deck arch.

2-165. Dry Canyon Creek Bridge, Rowena Vic., Wasco County, Oregon, 1921. Conde B. McCullough, designer; Whitman & Kuckenberg (Portland, OR), builder. Jet T. Lowe, photographer, 1990. P&P,HAER,ORE,33-ROW,1-1.

The Dry Canyon Creek Bridge is a 75-foot, reinforced concrete, ribbed deck arch. It is one of two noteworthy bridges McCullough designed on the Columbia River Highway. The other is the 110-foot Mosier Creek Bridge (1920).

2-164

2-165

2-166

2-166. Winchester Bridge, Umpqua River, Winchester, Douglas County, Oregon, 1922—24. Conde B. McCullough, designer; H. E. Doering (Portland, OR), builder. Jet T. Lowe, photographer, 1990. P&P,HAER,ORE, 10-WINC,1-1.

2-167. Pier, arch skewbacks, and spandrel columns, Winchester Bridge, Umpqua River, Winchester, Douglas County, Oregon, 1922–1924. Conde B. McCullough, designer; H. E. Doering (Portland, OR), builder. Jet T. Lowe, photographer, 1990. P&P,HAER,ORE,10-WINC,1-4.

2-167

2-168. Rogue River Bridge (Isaac Lee Patterson Memorial Bridge), Gold Beach, Curry County, Oregon, 1930–1932. Conde B. McCullough, designer; Mercer-Fraser Company (Eureka, CA), builder. Jet T. Lowe, photographer, 1990. P&P,HAER,ORE,8-GOBE,1-17.

This was the first bridge in the United States to use the system of prestressing reinforced concrete developed by the French engineer Eugène Freyssinet (1878–1943). The technique allowed McCullough to reduce the amount of concrete and reinforcing steel in the arches.

2-169. Details of Freyssinet method of concrete arch construction, Rogue River Bridge (Isaac Lee Patterson Memorial Bridge), Gold Beach, Curry County, Oregon, 1930–1932. Conde B. McCullough, designer; Mercer-Fraser Company (Eureka, CA), builder. Todd A. Croteau, Richard L. Koochagian, Gretchen Van Dusen, Rafael Villalobos, delineators, 1990. P&P,HAER,ORE,8-GOBE,1-, sheet no. 2.

Eugène Freyssinet's method involves constructing the arch in such a way as to carefully control the behavior of the concrete as loads are imposed. Before the crown is closed, powerful jacks compress the ribs, introducing stresses that (1) slightly change the shape of the arch so that it lifts up from the centering, which is then removed, and (2) offset the deformations that occur as the structure settles.

2-170. Layout of jacking apparatus, Rogue River Bridge (Isaac Lee Patterson Memorial Bridge), Gold Beach, Curry County, Oregon, 1930–1932. Conde B. McCullough, designer; Mercer-Fraser Company (Eureka, CA), builder. Todd A. Croteau, Richard L. Koochagian, Gretchen Van Dusen, Rafael Villalobos, delineators, 1990. P&P,HAER,ORE,8-GOBE,1-,sheet no. 4.

2-168

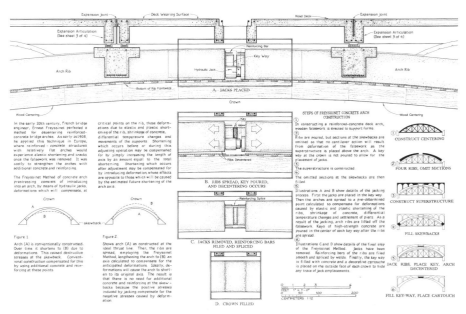

2-169

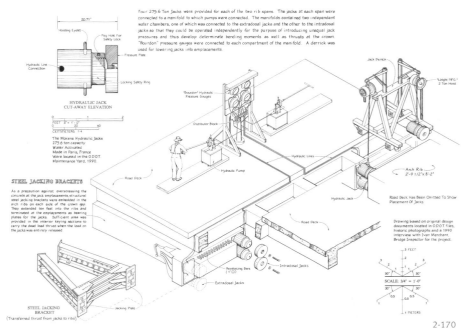

2-170

2-171

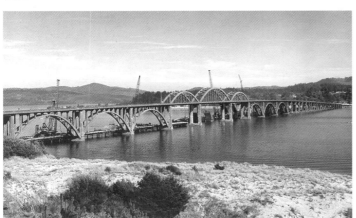

2-172

2-171. Cape Creek Bridge, Lane County, Oregon, 1931–1932. Conde B. McCullough, designer; John K. Holt (Salem, OR), builder, main arch; Clackamas Construction Company (Oregon City, OR), builder, north viaduct. Jet T. Lowe, photographer, 1990. P&P,HAER,ORE,20-FLO,1-1.

Poor foundation conditions in the streambed required the bridge's dead load to be distributed over many bents. McCullough addressed the problem by designing the bridge as a two-tiered viaduct with a simple arch spanning the main channel.

2-172. Alsea Bay Bridge, Waldport, Lincoln County, Oregon, 1934–1936 (replaced 1992). Conde B. McCullough, designer; Lindstrom & Feigenson, Park & Banfield, T. H. Banfield (Portland, OR), builders. Jet T. Lowe, photographer, 1990. P&P,HAER,ORE,21-WALPO,1-6.

McCullough's 3,028-foot Alsea Bay Bridge included reinforced concrete deck-girder and deck-arch approaches to three reinforced concrete through-tied arches over the bay's navigable channel. It was one of five large spans McCullough's engineers realized on the Oregon Coast Highway between 1933 and 1936. The ambitious enterprise, funded with assistance from the New Deal's Public Works Administration, provided jobs during the Great Depression.

2-173. Hinge details, Alsea Bay Bridge, Waldport, Lincoln County, Oregon, 1934–1936 (replaced 1992). Conde B. McCullough, designer; Lindstrom & Feigenson, Park & Banfield, T. H. Banfield (Portland, OR), builders. Todd A. Croteau, Richard L. Koochagian, Gretchen Van Dusen, Rafael Villalobos, delineators, 1990. P&P,HAER,ORE,21-WALPO,1-,sheet no. 2.

For this and a number of his other concrete arch bridges, McCullough employed three hinges. At the crown he used a Considère hinge, invented by the French engineer Armand Considère (1841–1914), which conveyed movement through the flexibility of the bunched reinforcing bars (known as a bowtie) rather than the traditional pinned pivot hinge.

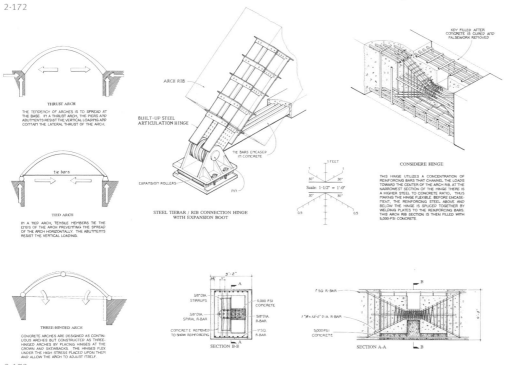

2-173

2-174

2-174. Rogue River Bridge (Caveman Bridge), Grant Pass, Josephine County, Oregon, 1931. Conde B. McCullough, designer. Jet T. Lowe, photographer, 1990. P&P,HAER,ORE,17-GRAPA,1-3.

2-175. View of deck, Rogue River Bridge (Caveman Bridge), Grant Pass, Josephine County, Oregon, 1931. Conde B. McCullough, designer. Jet T. Lowe, photographer, 1990. P&P,HAER,ORE,17-GRAPA,1-4.

2-175

TRUSS BRIDGES

IT IS NOT ALWAYS PRACTICAL TO FABRICATE MONOLITHIC BEAMS OR ARCHES with the depths required to sustain long spans and heavy loads. However, the structural properties of a beam or arch may also be achieved by a truss, an assembly of relatively short components arranged as a series of triangles. Like beams, trusses can be designed as simple or continuous spans and as cantilevers (structural elements extending beyond their point of support).

In the late eighteenth century, American bridge builders adopted and adapted truss patterns commonly used in medieval and early modern Europe for framing roofs and, at least since the Renaissance, for the construction of bridges. Throughout the nineteenth century, American designers experimented with various forms of trusses in wood, iron, and steel; they patented hundreds of configurations to try to make them proprietary technologies. Consolidation among manufacturers and preference for standardization by governmental engineering departments in the late nineteenth and early twentieth centuries led to the dominance of the Pratt, Parker, and Warren trusses, still common today.

In addition to the proportions and arrangement of its members, a critical issue in truss design is the detailing of connections, which can be flexible, thereby allowing local adjustments to stresses, or rigid, which requires all stresses to be resisted by the truss as a whole. The common use of pinned connections, around which the members of the joint can rotate, distinguished American bridge design from typical European practice in the nineteenth century. Rigid connections, achieved by bolts, rivets, or welding, became dominant in the twentieth century.

Bridges typically are built with a pair of trusses, one on each side of the deck. This arrangement may be classed in three categories: deck trusses support the roadway from below; through trusses contain the roadway between the truss panels, which rise high above the deck and are connected by overhead braces; pony or half-through trusses are shallower and without overhead braces. The structural members of wooden truss bridges usually are covered to protect them from direct exposure to the elements.

This section surveys the following truss types and offers a sampling of the many variations that occur within each category:

King-post truss (3-001–3-008): in the simplest configuration, a triangle bisected by a vertical post; the diagonal panels may be repeated symmetrically to form a multiple king-post truss.

Waddell "A" truss (3-009–3-010): patented by John Alexander Low Waddell in 1894; a steel truss that may be likened to a single king-post through truss with extensive cross bracing.

Queen-post truss (3-011–3-014): in the simplest configuration, a three-panel, parallel chord structure consisting of diagonal panels at the ends and a rectangular panel at the center.

Arch-reinforced truss (3-015–3-021): a structure in which an arch has been superimposed on the truss; nineteenth-century American bridge builders frequently employed it in timber construction.

Bowstring arch truss (3-022–3-077): commonly used for road bridges in the middle third of the nineteenth century; constructed of iron or steel (late examples), it consists of a curved or polygonal top chord and horizontal bottom chord (forming a tied arch) and panels with vertical posts and diagonal braces.

Suspension truss (3-078–3-084): a wrought-iron chain reinforces or replaces the bottom chord of the truss.

Lenticular truss (3-085–3-108): its iron or steel curved upper and lower chords resemble the shape of a lens.

Town lattice truss (3-109–3-122): patented by Ithiel Town in 1820, a diagonal lattice of uniform wooden members connecting the top and bottom chords.

Long truss (3-123–3-128): patented by Stephen Long in the 1830s; its timber panels have diagonal cross bracing in compression between vertical posts in tension.

Howe truss (3-129–3-156): a variant of the Long truss patented by William Howe in 1840 that replaces the wooden posts with wrought-iron rods; subsequently adapted to all-metal and, less frequently, reinforced concrete construction.

Pratt truss (3-157–3-217): named for Thomas and Caleb Pratt, who patented a composite wood and iron truss in 1844; superficially similar to the Howe truss in appearance, it distributes loads differently, placing the vertical posts in compression and the diagonals in tension; widely used in the nineteenth and twentieth centuries under many design patents.

Parker truss (3-218–3-234): variant of the Pratt truss patented by Charles H. Parker between 1868 and 1871 and built in many variations; the curved or polygonal top chord allows the truss to be deeper at the center of the span than at the ends.

Baltimore truss (3-235–3-243): a type of Pratt truss with subdivided panels; introduced by the Baltimore & Ohio Railroad in 1871.

Pennsylvania truss (3-244–3-263): also known as the Petit truss; developed by the Pennsylvania Railroad in 1875, it is a type of Pratt truss with subdivided panels and a polygonal upper chord.

Whipple-Murphy truss (3-264–3-295): based on patents obtained in the late 1840s by Squire Whipple and John W. Murphy and commonly used to the end of the nineteenth century; it is a Pratt truss with diagonal members extending across two panels.

Triple-intersection Pratt truss (3-296–3-297): introduced in the 1870s, a rarely-built truss with diagonals extending across three panels.

Bollman truss (3-298–3-304): developed by Wendel Bollman around 1850, it superimposes a suspension system consisting of long, wrought-iron bars extending from the end posts to each panel point upon the post-and-diagonal panel system of the Pratt truss.

Fink truss (3-305–3-314): a suspension and panel system designed by Albert Fink in the 1850s; similar to the Bollman truss, it has a simpler layout with symmetrical diagonals.

Post truss (3-315–3-319): invented by Simeon S. Post in the mid-1860s; both the tension and the compression members are inclined.

Thacher truss (3-320–3-322): combined panel and suspension truss patented by Edwin Thacher in 1883; rarely built.

K-truss (3-323): panel system used for steel-truss bridges in the twentieth century in which the webs (the members within the panels) have a K-shape arrangement formed by the intersection of a pair of diagonal braces at the midpoint of one post.

Warren truss (3-324–3-354): patented in England in 1848 by James Warren and Theobald Willoughby Monsani, it consists of diagonal members, alternately placed in tension and compression, forming a W pattern; widely used, especially in the twentieth century.

Cantilevered truss (3-355–3-390): a truss (of any configuration) built to extend beyond its point of support.

KING-POST TRUSS

The king-post truss has been in use since the Middle Ages. In its simplest configuration, it consists of a triangle bisected by a vertical tie, known as the post although it is a tension member. Additional vertical and diagonal braces may be added. For bridges, pairs of king-post trusses may be used singly for short spans (3-002), in a series of simple spans (3-001), and in an extended configuration—the multiple king-post truss (3-004–3-008).

3-001. New Fork River Bridge, Boulder Vic., Sublette County, Wyoming, 1917. Lincoln County, builder. Clayton B. Fraser, photographer, 1982. P&P,HAER,WYO,18-BOUL.V,1-3.

The bridge consists of two 46-foot timber king-post pony trusses. They have steel rods as the king-posts and numerous braces.

3-001

3-002. Fourteenth Street Bridge, spanning Missouri-Pacific Railroad tracks, North Little Rock, Arkansas, 1925 (demolished 1988). Michael Swanda, photographer, 1988. P&P,HAER,ARK,60-NOLI,4-2.

This timber king-post pony truss had vertical braces and a steel king-post.

3-003. South Fork Malheur River Bridge (No. 25E33), Crane Vic., Harney County, Oregon, 1964 (demolished ca. 1998). Harney County, builder. James Norman, photographer, 1996. P&P,HAER,ORE,13-CRAN.V,1-5.

Riveted steel plates (gussets) and timber braces reinforced this 39-foot timber king-post through truss.

3-004. Humpback Covered Bridge, Dunlap Creek, Covington Vic., Alleghany County, Virginia, 1857. Thomas McDowell Kincaid, builder. Jack E. Boucher, photographer, 1971. P&P,HAER,VA,3-COV.V,1-6.

This 100-foot wooden span is the third bridge on this site. It was built as part of the James River & Kanawha Turnpike, which linked the coastal market centers with settlements west of the Allegheny Mountains.

3-002

3-003

3-004

3-005

3-006

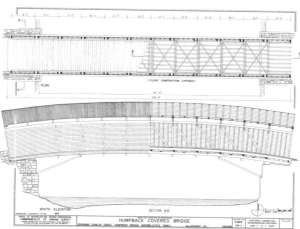

3-007

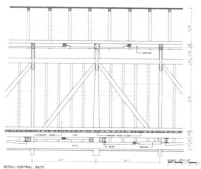

3-008

3-005. Interior view, Humpback Covered Bridge, Dunlap Creek, Covington Vic., Alleghany County, Virginia, 1857. Thomas McDowell Kincaid, builder. Jack E. Boucher, photographer, 1971. P&P,HAER,VA,3-COV.V,1-13.

The triangular king-post truss appears in its purest form at the center of the span, from which it is then extended symmetrically as a multiple king-post truss.

3-006. View of deck structure from below, Humpback Covered Bridge, Dunlap Creek, Covington Vic., Alleghany County, Virginia, 1857. Thomas McDowell Kincaid, builder. Jack E. Boucher, photographer, 1971. P&P,HAER,VA,3-COV.V,1-9.

3-007. Elevations and plan, Humpback Covered Bridge, Covington Vic., Alleghany County, Virginia, 1857. Thomas McDowell Kincaid, builder. Charles King, delineator, 1970. P&P,HAER,VA,3-COV.V,1-,sheet no. 2.

The builders curved the upper and lower chords of the truss to impart a degree of arched action that they likely intended to stiffen the structure.

3-008. Detail of king-post truss at central panels, Humpback Covered Bridge, Dunlap Creek, Covington Vic., Alleghany County, Virginia, 1857. Thomas McDowell Kincaid, builder. Donald G. Prycer, delineator, ca. 1971. P&P,HAER,VA,3-COV.V,1-,sheet no. 4.

WADDELL "A" TRUSS

John Alexander Low Waddell (1854–1938) was among the most prominent bridge design-ers and authors on bridge engineering in the late nineteenth and early twentieth centuries. He created the "A" truss (patented in 1894) as an economical solution for pin-connected steel railroad bridges with spans in the range of 100 feet and greater rigidity than stan-dard Pratt truss configurations then available. The design may be likened to a single king-post through truss with extensive cross-bracing and vertical and diagonal braces in the webs. It was employed by railroad companies in the midwestern United States and by the Nippon Railway Company in Japan until new, more rigid versions of the Pratt truss became available in the first decade of the twentieth century.

3-009. Waddell "A" Truss Bridge, English Landing Park, Parkville, Platte County, Missouri (original location Linn Branch Creek, Trimble Vic., Clinton County, MO), 1898. J. A. L. Waddell, engineer; A & P Roberts Company and Pencoyd Bridge Company (Pencoyd, PA), builders. Unidentified photographer, ca. 1980. P&P,HAER, MO,25-TRIM.V,1-8.

Erected for the Quincy, Omaha, & Kansas City Railway, this bridge was converted into a highway bridge in 1953, dismantled in 1980, and reconstructed as a pedestrian bridge.

3-010. Details of pin connections, Waddell "A" Truss Bridge, Eng-lish Landing Park, Parkville, Platte County, Missouri (original loca-tion Linn Branch Creek, Trimble Vic., Clinton County, MO), 1898. J. A. L. Waddell, engineer; A & P Roberts Company and Pencoyd Bridge Company (Pencoyd, PA), builders. Richard K. Andersen, delineator, 1980. P&P,HAER,MO,25-TRIM.V,1-,sheet no. 2.

3-009

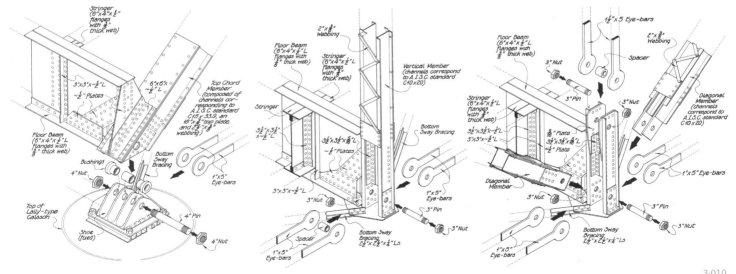

3-010

QUEEN-POST TRUSS

Rather than meeting to form the apex of a triangle, the diagonal members of the queen-post truss are separated by a rectangular panel. Like the king-post truss, queen-post trusses may have additional vertical and diagonal braces.

3-011

3-012

3-011. Zurich Road Bridge, spanning Southern Pacific, Chicago & St. Louis Railroad tracks, Joliet Vic., Will County, Illinois, ca. 1890. Southern Pacific Railroad, builder. Roger McCredie, photographer, 1995. P&P,HAER,ILL,99-JOL.V, 2-3.

3-012. Wawona Covered Bridge, South Fork Merced River, Yosemite National Park, Mariposa County, California, 1868 (covered 1878, rebuilt 1956, rehabilitated 2002–2004). Galen Clark, builder; covered by Henry Washburn. Brian C. Grogan, photographer, 1991. P&P,HAER,CAL,22-WAWO,5-12.

3-013. Interior, Wawona Covered Bridge, South Fork Merced River, Yosemite National Park, Mariposa County, California, 1868 (covered 1878, rebuilt 1956, rehabilitated 2002–2004). Galen Clark, builder; covered by Henry Washburn. Brian C. Grogan, photographer, 1991. P&P,HAER,CAL,22-WAWO,5-7.

3-014. Cutaway view, Wawona Covered Bridge, South Fork Merced River, Yosemite National Park, Mariposa County, California, 1868 (covered 1878, rebuilt 1956, rehabilitated 2002–2004). Galen Clark, builder; covered by Henry Washburn. Dione de Martelaere, delineator, 1991. P&P,HAER,CAL,22-WAWO,5-,sheet no. 2.

3-013

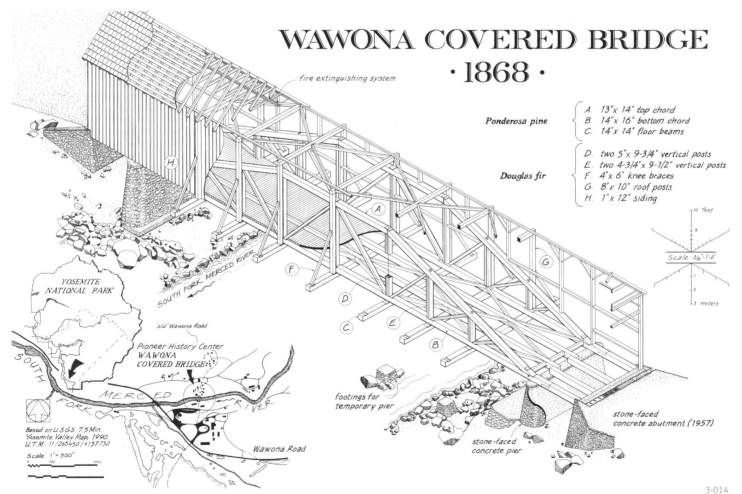

WAWONA COVERED BRIDGE
· 1868 ·

fire extinguishing system

Ponderosa pine
A. 13" x 14" top chord
B. 14" x 16" bottom chord
C. 14" x 14" floor beams

Douglas fir
D. two 5" x 9-3/4" vertical posts
E. two 4-3/4" x 9-1/2" vertical posts
F. 4" x 6" knee braces
G. 8" x 10" roof posts
H. 1" x 12" siding

Scale 3/16"=1'-0"

10 feet
5
1
2
3 meters

YOSEMITE NATIONAL PARK

SOUTH FORK MERCED RIVER

'old' Wawona Road

Pioneer History Center
WAWONA COVERED BRIDGE

SOUTH FORK MERCED RIVER

Wawona Road

Based on U.S.G.S. 7.5 Min.
Yosemite Valley Map, 1990.
U.T.M.: 11/263450/4157750

Scale 1"= 500'
0   500   1000

footings for temporary pier

stone-faced concrete pier

stone-faced concrete abutment (1957)

3-014

ARCH-REINFORCED TRUSS

During the late eighteenth and early nineteenth centuries, builders applied their empirical knowledge of arches and simple trusses to create composite structures that allowed them to span longer distances.

3-015

3-015. Permanent Bridge, Schuylkill River, Philadelphia, Pennsylvania, 1800–1805 (replaced 1850). Timothy Palmer, designer. William R. Birch, engraver, ca. 1807. P&P, 4384 G.

Timothy Palmer (1751–1821) patented a design combining a multiple king-post truss and an arch in 1797 and built several long-span bridges in New England before receiving the commission for this 550-foot bridge. Located at the foot of Market Street, it replaced a pontoon bridge and provided an all-weather crossing for the Lancaster-Philadelphia Turnpike. At the suggestion of Judge Richard Peters, president of the bridge company, Palmer enclosed the truss and created the first known covered bridge in the United States.

Burr Arch Truss

3-016. Covered Bridge, South Fork of Licking River, Cynthiana, Harrison County, Kentucky, 1837 (demolished 1946–1948). Greenup Remington, builder. Theodore Webb, photographer, 1934. P&P, HABS, KY, 49-CYNTH, 1-1.

3-017. Elevation and sections, Covered Bridge, South Fork of Licking River, Cynthiana, Harrison County, Kentucky, 1837 (demolished). Greenup Remington, builder. Chester H. Disque, delineator, 1934. P&P, HABS, KY, 49-CYNTH, 1, sheet no. 3.

This exemplifies the arch-reinforced, multiple king-post truss design patented by Theodore Burr (1771–1822) in 1817 and used widely for wooden bridges in the United States for most of the 19th century. In contrast to Palmer's design (3-019), Burr's system offered a more efficient integration of arch and truss and allowed the roadway to be level.

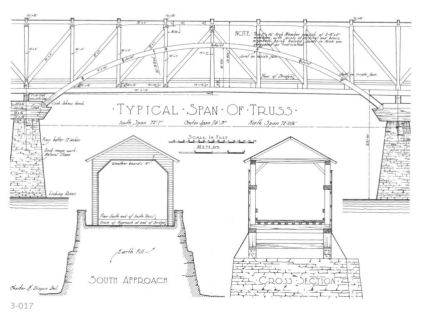

3-016

3-017

3-018

3-019

3-018. Griesemer's Mill Covered Bridge, Manatawny Creek, Yellow House Vic., Berks County, Pennsylvania, 1832. Cervin Robinson, photographer, 1958. P&P,HABS,PA,6-YEL.V,2-1.

3-019. Interior, Griesemer's Mill Covered Bridge, Manatawny Creek, Yellow House Vic., Berks County, Pennsylvania, 1832. Cervin Robinson, photographer, 1958. P&P,HABS,PA,6-YEL.V,2-2.

3-020. Barrackville Covered Bridge, Buffalo Creek, Barrackville, Marion County, West Virginia, 1853. Lemuel and Eli Chenoweth, builders. William E. Barrett, photographer, 1978. P&P,HAER,WVA,25-BARAC,1-7.

Lemuel (1811–1887) and Eli Chenoweth were prominent carpenters who built a number of bridges at mid-century as part of Virginia's (now West Virginia's) turnpike system.

3-021. Interior viewed from sidewalk, Barrackville Covered Bridge, Buffalo Creek, Barrackville, Marion County, West Virginia, 1853. Lemuel and Eli Chenoweth, builders. William E. Barrett, photographer, 1978. P&P,HAER, WVA,25-BARAC,1-9.

The iron rods from the arch to the bottom chord were added in 1934.

3-020

3-021

BOWSTRING ARCH TRUSS

Bridge designers have erected many configurations combining the arch and the trangulated truss. Squire Whipple (1804–1888) obtained a patent for the design of a bowstring arch truss in 1841. It features a polygonal top chord, horizontal bottom chord, vertical posts, and diagonal braces. Economical in its use of materials and structurally efficient, it was particularly well suited for canal bridges and became very popular with builders, who patented many variations through the 1870s. Much attention was directed to the composition of the upper chord, and designers experimented with various arrangements of iron tubes, cruciform girders, and plates, seeking technological efficiency and profit from license fees. The necessity of customizing the members of the upper chord to conform to the specific arc of each span eventually became a drawback compared to the more easily standardized components of the parallel-chord Pratt truss.

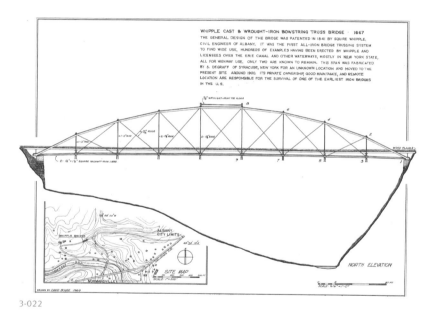

3-022

3-022. Elevation and site plan, Whipple Cast & Wrought Iron Bowstring Truss Bridge, spanning ravine near Normans Kill, Albany, Albany County, New York, 1867 (moved from unknown site to present, location ca. 1900). Squire Whipple, patent holder; Simon DeGraff (Syracuse, NY), builder. David Bouse, delineator, 1969. P&P,HAER,NY,1-ALB,19-,sheet no. 2.

DeGraff's 110-foot bridge closely follows Whipple's patented design of 1841. The upper chords are cast iron. The lower chords and the vertical and diagonal rods are wrought iron.

3-023. Whipple Cast & Wrought Iron Bowstring Truss Bridge, spanning ravine near Normans Kill, Albany, Albany County, New York, 1867 (moved from unknown site to present location ca. 1900). Squire Whipple, patent holder; Simon DeGraff (Syracuse, NY), builder. Jack E. Boucher, photographer, 1969. P&P,HAER, NY,1-ALB,19-4.

3-023

3-024

3-025

3-024. Chord connections at abutment, Whipple Cast & Wrought Iron Bowstring Truss Bridge, spanning ravine near Normans Kill, Albany, Albany
County, New York, 1867 (moved from unknown site to present location ca.
1900). Squire Whipple, patent holder; Simon DeGraff (Syracuse, NY), builder.
Jack E. Boucher, photographer, 1969. P&P,HAER,NY,1-ALB,19-6.

3-025. Upper chord and diagonal connections, Whipple Cast & Wrought Iron
Bowstring Truss Bridge, spanning ravine near Normans Kill, Albany, Albany
County, New York, 1867 (moved from unknown site to present location ca.
1900). Squire Whipple, patent holder; Simon DeGraff (Syracuse, NY), builder.
Jack E. Boucher, photographer, 1969. P&P,HAER,NY,1-ALB,19-7.

3-026. Bottom chord and deck beam connections, Whipple Cast & Wrought
Iron Bowstring Truss Bridge, spanning ravine near Normans Kill, Albany,
Albany County, New York, 1867 (moved from unknown site to present location
ca. 1900). Squire Whipple, patent holder; Simon DeGraff (Syracuse, NY),
builder. Jack E. Boucher, photographer, 1969. P&P,HAER,NY,1-ALB,19-8.

3-026

3-027

3-028

3-029

3-027. Squire Whipple Bridge (formerly Cayadutta Creek Bridge), Erie Canal, Vischer Ferry Nature & Historic Preserve, Town of Clifton Park, Saratoga County, New York ca. 1869 (moved to present location from Fonda, Montgomery County, NY, 1998). Squire Whipple, designer. Jet T. Lowe, photographer, 1994. P&P,HAER,NY,29-FOND,2-3.

3-028. Blackhoof Street Bridge, Miami-Erie Canal, New Bremen, Auglaize County, Ohio, 1864 (originally spanned Auglaize River in Wapakoneta, OH; moved in 1894 and again in 1984 to present location). David H. Morrison, Columbia Bridge Works (Dayton, OH), builder. Joseph Elliott, photographer, 1992. P&P,HAER,OHIO,6-NEWBR,1-2.

The 57-foot Blackhoof Bridge has bolted cast-iron compression members (upper chord and posts) and wrought-iron tension members (bottom chord and diagonals).

3-029. Detail of panels, Blackhoof Street Bridge, Miami-Erie Canal, New Bremen, Auglaize County, Ohio, 1864 (originally spanned Auglaize River in Wapakoneta, OH; moved in 1894 and again in 1984 to present location). David H. Morrison, Columbia Bridge Works (Dayton, OH), builder. Joseph Elliott, photographer, 1992. P&P,HAER,OHIO,6-NEWBR,1-6.

3-030. Upper Pacific Mills Bridge (Moseley Bridge), Merrimack College, North Andover, Essex County, Massachusetts, 1864 (photographed at original site spanning North Canal, Lawrence, MA; moved 1989). Thomas Moseley, Moseley Iron Building Works (Boston, MA), designer and builder. Martin Stupich, photographer, 1987. P&P,HAER,MASS,5-LAWR,6-1.

This 96-foot bridge is an example of the bowstring arch truss that Thomas Moseley (1813–1880) introduced in 1855 and continued to refine over the next twenty years. He fabricated its distinctive upper chord as a triangular tube composed of riveted wrought-iron boilerplate.

3-031. Detail of panels, Upper Pacific Mills Bridge (Moseley Bridge), Merrimack College, North Andover, Essex County, Massachusetts, 1864 (photographed at original site spanning North Canal, Lawrence, MA; moved 1989). Thomas Moseley, Moseley Iron Building Works (Boston, MA), designer and builder. Martin Stupich, photographer, 1987. P&P,HAER,MASS,5-LAWR,6-4.

3-032. Connection details, Upper Pacific Mills Bridge (Moseley Bridge), Merrimack College, North Andover, Essex County, Massachusetts, 1864 (moved to present site from North Canal, Lawrence, MA, 1989). Thomas Moseley, Moseley Iron Building Works (Boston, MA), designer and builder. Wayne Chang, delineator, 1987. P&P,HAER,MASS,5-LAWR,6-,sheet no. 2.

3-033. Connections at abutment, Upper Pacific Mills Bridge (Moseley Bridge), Merrimack College, North Andover, Essex County, Massachusetts, 1864 (moved to present site from North Canal, Lawrence, MA, 1989, and restored). Thomas Moseley, Moseley Iron Building Works (Boston, MA), designer and builder. Joseph Elliott, photographer, 1991. P&P,HAER,MASS,5-LAWR,6-14.

3-030

3-031

3-033

CONNECTION DETAILS

3-032

3-334.

3-334. Hares Hill Road Bridge, French Creek, Kimberton, Chester County, Pennsylvania, 1869. Thomas Moseley, Moseley Iron Bridge & Roof Company (Boston, MA), design and fabrication. Joseph Elliott, photographer, 1991. P&P,HAER,PA,15-KIMB,2-2.

Thomas Moseley developed a number of designs utilizing wrought iron during the 1850s and 1860s. This 103-foot bowstring arch truss with a wrought-iron lattice web conforms to a patent he received in 1869.

3-335. Interior, Hares Hill Road Bridge, French Creek, Kimberton, Chester County, Pennsylvania, 1869. Thomas Moseley, Moseley Iron Bridge & Roof Company (Boston, MA), design and fabrication. Joseph Elliott, photographer, 1991. P&P,HAER,PA,15-KIMB,2-6.

3-336. Detail of lattice and riveted connections, Hares Hill Road Bridge, French Creek, Kimberton, Chester County, Pennsylvania, 1869. Thomas Moseley, Moseley Iron Bridge & Roof Company (Boston, MA), design and fabrication. Joseph Elliott, photographer, 1991. P&P,HAER,PA,15-KIMB,2-10.

3-337. Abutment, Hares Hill Road Bridge, French Creek, Kimberton, Chester County, Pennsylvania, 1869. Thomas Moseley, Moseley Iron Bridge & Roof Company (Boston, MA), design and fabrication. Joseph Elliott, photographer, 1991. P&P,HAER,PA,15-KIMB,2-11.

3-335.

3-336.

3-337.

3-038. Henszey's Wrought-Iron Arch Bridge (Mosser's Bridge), Central Pennsylvania College, Summerdale, Cumberland County, Pennsylvania, 1869 (originally part of a two-span bridge in Slatington, PA; moved to Ontelaunee Creek, Lynnport Vic., Lehigh County, PA, 1900; moved to present location and rehabilitated, 2002). Joseph G. Henszey, designer; Daniel Beidelman, builder; Continental Bridge Company (Philadelphia, PA), fabricator. Joseph Elliott, photographer, 1991. P&P,HAER,PA,39-WANA,1-16.

Joseph G. Henszey was a hardware merchant turned bridge designer. This bridge employs features he patented in 1869, most notably the construction of the top chord with Phoenix column segments—a system of rolled, flanged channels joined by rivets patented by the Phoenix Iron Works in 1862 (3-244–3-245).

3-039. Kern Bridge, LeSueur River, Skyline Vic., Blue Earth County, Minnesota, 1873. Wrought Iron Bridge Company (Canton, OH), builder. Jet T. Lowe, photographer, 1990. P&P,HAER,MINN,7-SKY.V,1-3.

3-040. Interior view, Kern Bridge, LeSueur River, Skyline Vic., Blue Earth County, Minnesota, 1873. Wrought Iron Bridge Company (Canton, OH), builder. Jet T. Lowe, photographer, 1990. P&P,HAER,MINN,7-SKY.V,1-6.

3-041

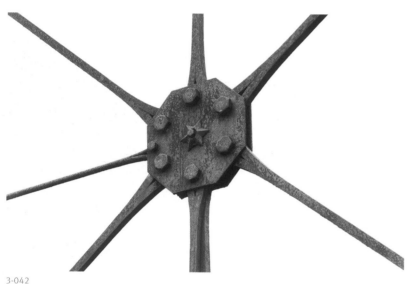

3-042

3-041. Freeport Bridge, Trout Run Park, Decorah, Winneshiek County, Iowa, 1878 (photographed at original site spanning Upper Iowa River near Freeport, IA; moved after 1985). Wrought Iron Bridge Company (Canton, OH), builder. Clayton B. Fraser, photographer, 1985. P&P,HAER,IOWA,96-DECOR.V,1-10.

3-042. Diagonal connection, Freeport Bridge, Trout Run Park, Decorah, Winneshiek County, Iowa, 1878 (photographed at original site spanning Upper Iowa River near Freeport, IA; moved after 1985). Wrought Iron Bridge Company (Canton, OH), builder. Clayton B. Fraser, photographer, 1985. P&P,HAER,IOWA,96-DECOR.V, 1-34.

3-043. Connection details, Freeport Bridge, Trout Run Park, Decorah, Winneshiek County, Iowa, 1878 (moved from original site spanning Upper Iowa River near Freeport, IA, after 1985). Wrought Iron Bridge Company (Canton, OH), builder. Clayton B. Fraser, delineator, 1985. P&P,HAER,IOWA,96-DECOR.V,1-,sheet no. 3.

Like many other bridge companies, the Wrought Iron Bridge Company developed or acquired proprietary components for its truss assemblies. In this and the previous example, the wrought-iron upper chord is a tubular arch composed of four shaped plates and a flat stiffening plate riveted together.

3-044. Panel detail, Freeport Bridge, Trout Run Park, Decorah, Winneshiek County, Iowa, 1878 (photographed at original site spanning Upper Iowa River near Freeport, IA; moved after 1985). Wrought Iron Bridge Company (Canton, OH), builder. Clayton B. Fraser, photographer, 1985. P&P,HAER,IOWA,96-DECOR.V,1-21.

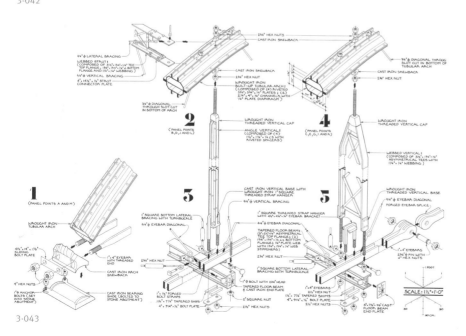

3-043

3-044

3-045

3-046

3-045. Deck connections, Freeport Bridge, Trout Run Park, Decorah, Winneshiek County, Iowa, 1878 (photographed at original site spanning Upper Iowa River near Freeport, IA; moved after 1985). Wrought Iron Bridge Company (Canton, OH), builder. Clayton B. Fraser, photographer, 1985. P&P,HAER,IOWA,96-DECOR.V,1-26.

3-046. Bottom chord, pinned eyebar connection, Freeport Bridge, Trout Run Park, Decorah, Winneshiek County, Iowa, 1878 (photographed at original site spanning Upper Iowa River near Freeport, IA; moved after 1985). Wrought Iron Bridge Company (Canton, OH), builder. Clayton B. Fraser, photographer, 1985. P&P,HAER,IOWA,96-DECOR.V,1-31.

3-047. Corbett's Mill Bridge, Marquoketa River, Scotch Grove Vic., Jones County, Iowa, 1871. Mahlon Miller and William Jamieson (Cleveland, OH), builders. Joseph Elliott, photographer, 1995. P&P,HAER,IOWA,53-SCOG.V,1-3.

Mahlon Miller received a patent for his tubular arch design in 1870.

3-048. Top chord composed of wrought-iron rods, Eureka Bridge, City Park, Castalia, Winneshiek County, Iowa, ca. 1872 (photographed at original site spanning Yellow River, Frankville Vic.; moved after 1983). Allen, McEvoy & Company (Beloit, WI), builder. J Ceronie, Dennett, Muessig, Ryan & Associates, photographer, 1983. P&P,HAER,IOWA,96-FRANK.V,1-10.

3-047

3-048

3-049

3-050

3-049. Eureka Bridge, City Park, Castalia, Winneshiek County, Iowa, ca. 1872 (photographed at original site spanning Yellow River, Frankville Vic.; moved after 1983). Allen, McEvoy & Company (Beloit, WI), builder. J Ceronie, Dennett, Muessig, Ryan & Associates, photographer, 1983. P&P,HAER,IOWA,96-FRANK.V,1-5.

3-050. Underside of deck, Eureka Bridge, City Park, Castalia, Winneshiek County, Iowa, ca. 1872 (photographed at original site spanning Yellow River, Frankville Vic.; moved after 1983). Allen, McEvoy & Company (Beloit, WI), builder. J Ceronie, Dennett, Muessig, Ryan & Associates, photographer, 1983. P&P,HAER,IOWA,96-FRANK.V,1-7.

3-051. Tioronda Bridge, Fishkill Creek, Beacon, Dutchess County, New York, 1872–1873. John Glass, George P. Schneider, William B. Rezner, Ohio Bridge Company (Cleveland, OH), designers and builder. Martin Stupich, photographer, 1987. P&P,HAER,NY,14-BEAC,3-1.

3-051

3-052

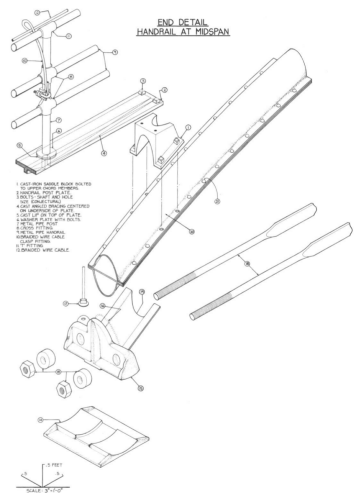

3-052. Bowstring truss, Tioronda Bridge, Fishkill Creek, Beacon, Dutchess County, New York, 1872–1873. John Glass, George P. Schneider, William B. Rezner, Ohio Bridge Company (Cleveland, OH), designers and builder. Martin Stupich, photographer, 1987. P&P,HAER,NY,14-BEAC,3-11.

The upper chords are composed of two sections of wrought iron forming an ellipse and a stiffening plate.

3-053. Foot-block patented by William B. Rezner, Tioronda Bridge, Fishkill Creek, Beacon, Dutchess County, New York, 1872–1873. John Glass, George P. Schneider, William B. Rezner, Ohio Bridge Company (Cleveland, OH), designers and builder. Martin Stupich, photographer, 1987. P&P,HAER,NY, 14-BEAC,3-13.

William B. Rezner received a patent for this cast-iron foot-block in 1872.

3-054. Connection details of handrail and foot-block on the skewback, Tioronda Bridge, Fishkill Creek, Beacon, Dutchess County, New York, 1872–1873. John Glass, George P. Schneider, William B. Rezner, Ohio Bridge Company (Cleveland, OH), designers and builder. Kim Kuykendall and Donald M. Durst, delineators, 1987. P&P,HAER,NY,14-BEAC,3-,sheet no. 3.

Rezner's foot-block receives the members of the top and bottom chords and allows the assembly to rotate as the structure expands and contracts.

END DETAIL
HANDRAIL AT MIDSPAN

1. CAST-IRON SADDLE BLOCK BOLTED TO UPPER CHORD MEMBERS.
2. HANDRAIL POST PLATE.
3. BOLTS - SHAFT AND HOLE SIZE (CONJECTURAL)
4. CAST ANGLED BRACING CENTERED ON UNDERSIDE OF PLATE.
5. CAST LIP ON TOP OF PLATE.
6. WASHER PLATE WITH BOLTS.
7. METAL PIPE POST
8. CROSS FITTING
9. METAL PIPE HANDRAIL.
10. BRAIDED WIRE CABLE CLASP FITTING.
11. "T" FITTING
12. BRAIDED WIRE CABLE

SCALE: 3"=1'-0"

3-054

3-055

3-056

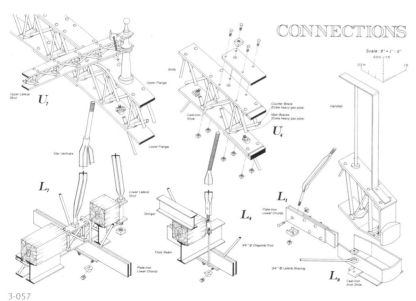

3-057

3-058

3-055. Fremont Mill Bridge (Central Park Bridge), Central Park, Center Junction, Anamosa Vic., Jones County, Iowa, 1873 (originally spanned Maquoketa River, Monticello). Joseph Davenport, Massillon Iron Bridge Company (Massillon, OH), designer and builder. Joseph Elliott, photographer, 1995. P&P,HAER,IOWA,53-ANAM.V,1-2.

3-056. Detail of top chord, Fremont Mill Bridge (Central Park Bridge), Central Park, Center Junction, Anamosa Vic., Jones County, Iowa, 1873 (originally spanned Maquoketa River, Monticello). Joseph Davenport, Massillon Iron Bridge Company (Massillon, OH), designer and builder. Joseph Elliott, photographer, 1995. P&P,HAER,IOWA,53-ANAM.V,1-7.

Joseph Davenport designed the upper chord as an arch formed by a Howe truss (for another example of his use of the Howe truss, see 3-150).

3-057. Connection details, Fremont Mill Bridge (Central Park Bridge), Central Park, Center Junction, Anamosa Vic., Jones County, Iowa, 1873 (originally spanned Maquoketa River, Monticello). Joseph Davenport, Massillon Iron Bridge Company (Massillon, OH), designer and builder. Roger Chien, delineator, 1995. P&P,HAER,IOWA,53-ANAM.V,1-,sheet no. 3.

3-058. Connections of lateral brace at top chord, Fremont Mill Bridge (Central Park Bridge), Central Park, Center Junction, Anamosa Vic., Jones County, Iowa, 1873 (originally spanned Maquoketa River, Monticello). Joseph Davenport, Massillon Iron Bridge Company (Massillon, OH), designer and builder. Joseph Elliott, photographer, 1995. P&P,HAER,IOWA,53-ANAM.V,1-6.

3-059. Fort Laramie Army Bridge (North Platte River bowstring truss bridge), Fort Laramie Vic., Goshen County, Wyoming, 1875. King Iron Bridge & Manufacturing Company (Cleveland, OH), builder. Jack E. Boucher, photographer, 1974. P&P,HAER,WYO,8-FOLA.V,1-1.

3-060. Interior view, Fort Laramie Army Bridge (North Platte River bowstring truss bridge), Fort Laramie Vic., Goshen County, Wyoming, 1875. King Iron Bridge & Manufacturing Company (Cleveland, OH), builder. Jack E. Boucher, photographer, 1974. P&P,HAER,WYO,8-FOLA.V,1-4.

3-061. Pier and truss connections, Fort Laramie Army Bridge (North Platte River bowstring truss bridge), Fort Laramie Vic., Goshen County, Wyoming, 1875. King Iron Bridge & Manufacturing Company (Cleveland, OH), builder. Jack E. Boucher, photographer, 1974. P&P,HAER,WYO,8-FOLA.V,1-16.

3-062

3-062. Egypt Pike Bridge, Mud Run, New Holland, Pickaway County, Ohio, 1876. Jonathan and Zimri Wall, Champion Iron Bridge Company (Wilmington, OH), designers and builder. Joseph Elliott, photographer, 1992. P&P,HAER,OHIO,65-NEWHO,1-5.

The Walls patented the Champion Wrought Iron Arch Bridge in 1874. Its top chord consists of three concentric arches of wrought-iron plates connected by a system of bolted blocks, tubes, and threaded rods.

3-063. Connection details, Egypt Pike Bridge, Mud Run, New Holland, Pickaway County, Ohio, 1876. Jonathan and Zimri Wall, Champion Iron Bridge Company (Wilmington, OH), designers and builder. Joseph A. Boquiren and Christina L. Madrid, delineators, 1992–1994. P&P,HAER,OHIO,65-NEWHO, 1-,sheet no. 2.

3-064. Roaring Run Bridge, I-81 Ironto Wayside, Montgomery County Virginia, 1877–1878 (originally spanned Stoney Fork, near Moneta; moved ca. 1930 to Roaring Run, Bedford County; restored and moved to present location, 1978). King Iron Bridge & Manufacturing Company (Cleveland, OH), builder. Jack E. Boucher, photographer, 1971. P&P,HAER, VA,10-BED.V,2-1.

3-065. Exploded isometric view of main joints, Roaring Run Bridge, Bedford Vic., I-81 Ironto Wayside, Montgomery County, Virginia, 1877–1878. King Iron Bridge & Manufacturing Company (Cleveland, OH), builder. Charles King, delineator, 1970. P&P, HAER,VA,10-BED.V,2-,sheet no. 3.

The top chords are composed of two riveted wrought-iron channels.

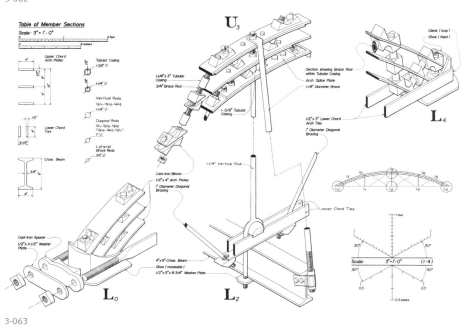

3-063

3-064

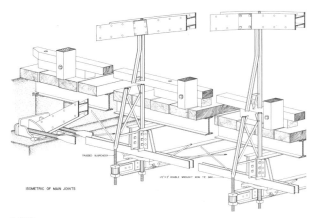

3-065

3-066. Tivoli Island Bridge, Rock River, Watertown, Jefferson County, Wisconsin, ca. 1877 (originally one of a four-span bridge crossing the Crawfish River, Milford; moved 1906). King Iron Bridge & Manufacturing Company (Cleveland, OH), builder. Martin Stupich, photographer, 1987. P&P,HAER,WIS,28-WATO,5-4.

3-067. Whetstone Creek Bridge, Mount Gilead, Morrow County, Ohio, 1879. Wrought Iron Bridge Company (Canton, OH), builder. Joseph Elliott, photographer, 1992. P&P,HAER,OHIO,59-MTGIL, 2-5.

The top chords are octagonal tubes composed of rolled, riveted iron plates, an assembly patented by William Laird of Canton in 1874.

3-068. Connection details, Whetstone Creek Bridge, Mount Gilead, Morrow County, Ohio, 1879. Wrought Iron Bridge Company (Canton, OH), builder. Troy Zimmermann, Elaine Pierce, Daron Fender, delineators, 1993. P&P,HAER,OHIO,59-MTGIL, 2-,sheet no. 3.

3-066

3-067

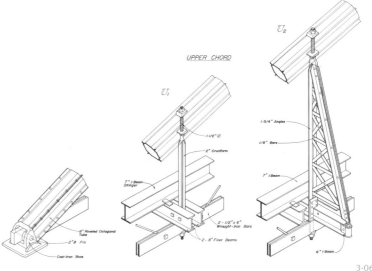

3-068

3-069

3-070

CONNECTION DETAILS

TOP CONNECTION

U_3

U_3

L_0

L_3

3/4" Rod

5" X 8" Plate

6 3/4" O Upper Chord (Phoenix Column)

3/4" O Lower Lateral Diagonal Bracing Rods

These Two Edges are not Consistently Flush Between Casting and Column

4" Square Nuts

3/4" X 1 1/4" hex nuts

4 1/8" Ø Washer

Cast-Iron Shoe

4" X 1 1/4" Plate

L_0

END TRUSS

POST

1/4" X 1 3/4" Angles

1" Bars

5/8" Rivets

Modern Bracket for Lateral Bracing

Brace Welded to Bracket

Bracket Welded to Lower Chord

4" RailRoad Rail Stringer

1 1/8" Ø Diagonal Bracing

1/4" X 2 1/4" Hex Nuts

4" X 7 1/4" Cast-Iron Plate with screws

5/8" Rivets

1" X 1 3/4" Hex Nuts

Tie Down

9" X 3 1/2" Traverse Beam

2 - 4" X 5/8" Lower Chords

3" X 8" Plates

FT 1   30 CM

FT 1

FT 1

CM 30   20   30 CM

Scale 3"x1'-0"

L_3

BOTTOM CONNECTION

5" X 8 1/2" Plates

5" X 8" Plates

3-071

3-069. Cooper's Tubular Arch Bridge, Old Erie Canal, De Witt, Onondaga County, New York, 1886 (originally spanned Canajoharie Creek in Canajoharie; moved 1975). William B. Cooper, engineer; Melvin A. Nash (Fort Edward, NY), builder. Jet T. Lowe, photographer, 1994. P&P,HAER,NY,34-DEWI,1-4.

3-070. Connection details, Cooper's Tubular Arch Bridge, Old Erie Canal, De Witt, Onondaga County, New York, 1886 (originally spanned Canajoharie Creek in Canajoharie; moved 1975). William B. Cooper, engineer; Melvin A. Nash (Fort Edward, NY), builder. Jet T. Lowe, photographer, 1994. P&P,HAER,NY,34-DEWI,1-9.

3-071. Connection details, Cooper's Tubular Arch Bridge, Old Erie Canal, De Witt, Onondaga County, New York, 1886 (originally spanned Canajoharie Creek in Canajoharie; moved 1975). William B. Cooper, engineer; Melvin A. Nash (Fort Edward, NY), builder. Luis G. Rosario-Lluveras, delineator, 1995. P&P,HAER,NY,34-DEWI,1-,sheet no. 3.

3-072

3-072. McGilvray Road Bridge No. 1, tributary of Black River, Van Loon Wildlife Area, La Crosse Vic., La Crosse County, Wisconsin, 1906. Charles M. Horton (Duluth, MN), engineer; La Crosse Bridge & Steel Company (LaCrosse, WI), builder. Martin Stupich, photographer, 1987. P&P,HAER,WIS,32-LACR.V,1-1.

This 134-foot-long bridge employs Charles M. Horton's system of clips and pins, which he patented in 1897 as an improvement on riveted connections. The arches of the top chords are composed of I-beam segments.

3-073. Connection details, McGilvray Road Bridge No. 1, tributary of Black River, Van Loon Wildlife Area, La Crosse Vic., La Crosse County, Wisconsin, 1906. Charles M. Horton (Duluth, MN), designer; La Crosse Bridge & Steel Company (LaCrosse, WI), builder. Martin Stupich, photographer, 1987. P&P,HAER,WIS,32-LACR.V,1-2.

3-074. Connection details, McGilvray Road Bridge No. 1, tributary of Black River, Van Loon Wildlife Area, La Crosse Vic., La Crosse County, Wisconsin, 1906. Charles M. Horton (Duluth, MN), designer; La Crosse Bridge & Steel Company (LaCrosse, WI), builder. David E. Jamison, 1987, delineator. P&P,HAER,WIS,32-LACR.V,1-,sheet no. 3.

3-073

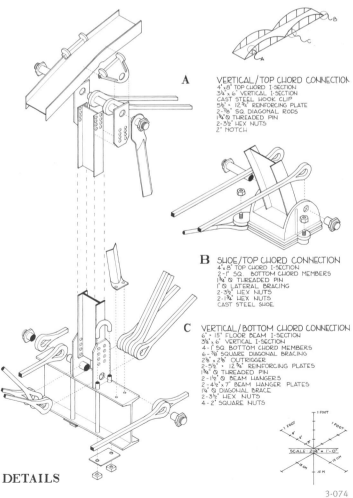

A VERTICAL/TOP CHORD CONNECTION
4"x8" TOP CHORD I-SECTION
3¼"x 6" VERTICAL I-SECTION
CAST STEEL HOOK CLIP
5½"x 12¾" REINFORCING PLATE
2-⅞" SQ. DIAGONAL RODS
1¾" ⌀ THREADED PIN
2-3½" HEX NUTS
2" NOTCH

B SHOE/TOP CHORD CONNECTION
4"x 8" TOP CHORD I-SECTION
2-1" SQ. BOTTOM CHORD MEMBERS
1¾" ⌀ THREADED PIN
1" ⌀ LATERAL BRACING
2-3½" HEX NUTS
2-1¾" HEX NUTS
CAST STEEL SHOE

C VERTICAL/BOTTOM CHORD CONNECTION
6"x 15" FLOOR BEAM I-SECTION
3¼"x 6" VERTICAL I-SECTION
4-1" SQ. BOTTOM CHORD MEMBERS
6-⅞" SQUARE DIAGONAL BRACING
2⅛" x 2⅛" OUTRIGGER
2-5½" x 12¾" REINFORCING PLATES
1¾" ⌀ THREADED PIN
2-1½" ⌀ BEAM HANGERS
2-4½"x 7" BEAM HANGER PLATES
1¼" ⌀ DIAGONAL BRACE
2-3½" HEX NUTS
4-2" SQUARE NUTS

SCALE: 2¼" = 1'-0"

DETAILS

3-074

3-075

3-076

3-075. Nishnabotna River Bridge, Manilla Vic., Crawford County, Iowa, 1945. Des Moines Steel Company, builder. Joseph Elliott, photographer, 1995. P&P,HAER,IOWA,24-MAN.V,1-1.

Sixty years after bowstring arch trusses had fallen out of common usage, engineers in Iowa grappling with domestic steel shortages during World War II readopted the type for short-span county highway bridges because of its structural efficiency.

3-076. Panel detail, Nishnabotna River Bridge, Manilla Vic., Crawford County, Iowa, 1945. Des Moines Steel Company, builder. Joseph Elliott, photographer, 1995. P&P,HAER,IOWA,24-MAN.V,1-3.

3-077. Detail of splice connection, Nishnabotna River Bridge, Manilla Vic., Crawford County, Iowa, 1945. Des Moines Steel Company, builder. Joseph Elliott, photographer, 1995. P&P,HAER,IOWA,24-MAN.V,1-6.

The 77-foot trusses were shop fabricated in two halves and assembled in the field using rivets and bolts. The truss members are built up steel angle iron, doubled for the chords and vertical members, single lengths for the diagonal members.

3-077

SUSPENSION TRUSS

Arches are in compression; inverted, they are in tension—as, for example a cable hanging between two points. Suspension trusses employ a wrought-iron eyebar chain that reinforces or replaces the bottom chord of the truss.

3-078. Germantown Covered Bridge, Little Twin Creek, Germantown, Montgomery County, Ohio, 1865 (not 1870). David H. Morrison, Columbia Bridge Works (Dayton, OH), designer and builder. Joseph Elliott, photographer, 1992. P&P,HAER,OHIO,57-GERM, 1-2.

3-079. Detail of suspension truss, Germantown Covered Bridge, Little Twin Creek, Germantown, Montgomery County, Ohio, 1865. David H. Morrison, Columbia Bridge Works (Dayton, OH), designer and builder. Joseph Elliott, photographer, 1992. P&P,HAER,OHIO, 57-GERM,1-6.

3-078

3-079

3-080

3-081

3-080. Detail of truss, John Bright No. 2 Covered Bridge, Ohio University Lancaster Campus, Lancaster, Fairfield County, Ohio, 1881 (originally spanned Poplar Creek, moved 1988). Augustus Borneman & Sons (Lancaster, OH), builder. Louise T. Cawood, photographer, 1986. P&P,HAER,OHIO,23-CAR.V,2-5.

The wooden reinforcing arch was added to the suspension truss in 1913.

3-081. John Bright No. 2 Covered Bridge, Ohio University Lancaster Campus, Lancaster, Fairfield County, Ohio, 1881 (originally spanned Poplar Creek, moved 1988). Augustus Borneman & Sons (Lancaster, OH), builder. Louise T. Cawood, photographer, 1986. P&P,HAER,OHIO,23-CAR.V,2-1.

3-082. Sections, John Bright No. 2 Covered Bridge, Ohio University Lancaster Campus, Lancaster, Fairfield County, Ohio, 1881 (originally spanned Poplar Creek, moved 1988). Augustus Borneman & Sons (Lancaster, OH), builder. Christopher P. Collins, delineator, 1986. P&P,HAER,OHIO,23-CAR.V,2-,sheet no. 1.

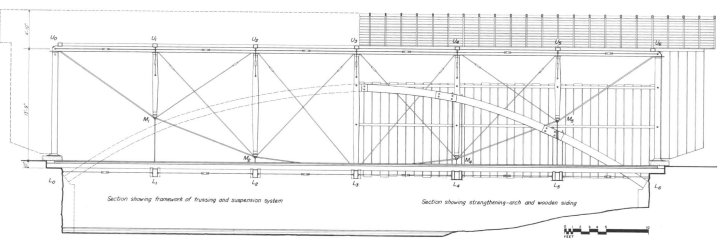

Section showing framework of trussing and suspension system

Section showing strengthening-arch and wooden siding

3-082

3-083. John Bright No. 1 Iron Bridge, Ohio University Lancaster Campus, Lancaster, Fairfield County, Ohio, ca. 1885 (originally spanned Poplar Creek, moved ca. 1988). Augustus Borneman, Hocking Valley Bridge Works (Lancaster, OH), builder. Louise T. Cawood, photographer, 1986. P&P,HAER, OHIO,23-CAR.V,1-1.

The combination of Pratt truss and eyebar chain is the same that Borneman used in the John Bright Bridge No. 2 (3-031–3-033), but here all the members are iron.

3-084. Elevation and plan, John Bright No. 1 Iron Bridge, Ohio University Lancaster Campus, Lancaster, Fairfield County, Ohio, ca. 1885 (originally spanned Poplar Creek, moved ca. 1988). Augustus Borneman, Hocking Valley Bridge Works (Lancaster, OH), builder. Christopher P. Collins, delineator, 1986. P&P,HAER,OHIO,23-CAR.V,1-, sheet no. 2.

3-083

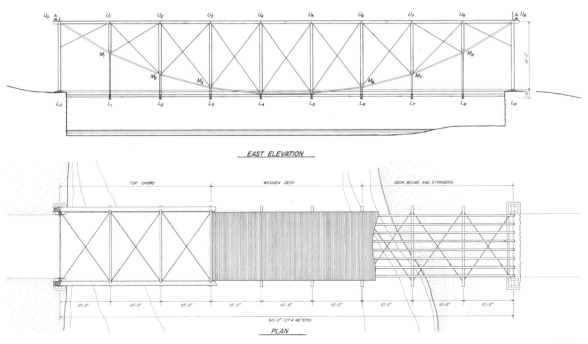

EAST ELEVATION

PLAN

3-084

The lenticular truss and its variant, named the Pauli truss after the German engineer August von Pauli (1802–1884), has curved upper and lower chords resembling the shape of a lens. Developed in England, Germany, and the United States in the mid-nineteenth century, the configuration allows the truss to be deeper in the middle than at the ends, thereby requiring less material than a Pratt truss of similar length.

3-085

3-085. Aerial view, Smithfield Street Bridge (at left), Monongahela River, Pittsburgh, Pennsylvania, 1883 (widened 1891). Gustav Lindenthal, engineer. Jack E. Boucher, photographer, 1974. P&P,HAER,PA,2-PITBU,58-7.

Hilly terrain and the confluence of the Allegheny and the Monongahela rivers forming the Ohio River have required the construction of approximately 2,000 bridges with spans of 8 feet or more in Pittsburgh and Allegheny County.

3-086. Smithfield Street Bridge, Monongahela River, Pittsburgh, Pennsylvania, 1883 (widened 1891; rehabilitated 1995). Gustav Lindenthal, engineer. Jack E. Boucher, photographer, 1974. P&P,HAER,PA,2-PITBU,58-30.

The Smithfield Street Bridge was the first major work by Gustav Lindenthal (see 2-073). Its two 236-foot lenticular trusses were the most ambitious application of the type in the United States at the time and are early examples of the use of steel in bridge construction.

3-086

3-087. Detail of trusses, Smithfield Street Bridge, Monongahela River, Pittsburgh, Pennsylvania, 1883 (widened 1891; rehabilitated 1995). Gustav Lindenthal, engineer. Jack E. Boucher, photographer, 1974. P&P,HAER,PA, 2-PITBU,58-26.

3-088. Original cast-iron portal, Smithfield Street Bridge, Monongahela River, Pittsburgh, Pennsylvania, 1883 (widened 1891). Gustav Lindenthal, engineer. Unidentified photographer, ca. 1883. P&P,LC-USZ62-80393.

3-089. View after widening, Smithfield Street Bridge, Monongahela River, Pittsburgh, Pennsylvania, 1883 (widened 1891). Gustav Lindenthal, engineer. DETR, photographer, ca. 1900. P&P,LC-D4-13891.

3-090. Steel portal, Smithfield Street Bridge, Monongahela River, Pittsburgh, Pennsylvania, 1883 (widened 1891, portal 1915; rehabilitated 1995). Gustav Lindenthal, engineer; Stanley L. Rouse, portal design. Jack E. Boucher, photographer, 1974. P&P,HAER,PA,2-PITBU,58-15.

3-091

3-092

3-093

3-091. Detail, Aiken Street Bridge, Merrimack River, Lowell, Middlesex County, Massachusetts, 1882–1883. Corrugated Metal Company (East Berlin, CT), builder. Martin Stupich, photographer, 1990. P&P,HAER,MASS,9-LOW,21-2.

The bridge consists of five 152-foot lenticular trusses based on a patent issued to William O. Douglas in 1878. The Corrugated Metal Company, later known as the Berlin Iron Bridge Company, purchased the rights to the patent and marketed a line of lenticular through and pony trusses. Its 1889 catalog listed 664 bridges in twelve states.

3-092. Interior view, Aiken Street Bridge, Merrimack River, Lowell, Middlesex County, Massachusetts, 1882–1883. Corrugated Metal Company (East Berlin, CT), builder. Martin Stupich, photographer, 1990. P&P,HAER,MASS,9-LOW, 21-4.

3-093. Tuttle Bridge, Housatonic River, Lee, Berkshire County, Massachusetts, 1885. Berlin Iron Bridge Company (East Berlin, CT), builder. Martin Stupich, photographer, 1990. P&P,HAER,MASS,2-LEE,1-6.

3-094. Old Corinth Road Bridge, Sacandaga River, Hadley, Saratoga County, New York, 1885 (rehabilitaed 2006). Berlin Iron Bridge Company (East Berlin, CT), builder. Jet T. Lowe, photographer, 1994. P&P,HAER,NY,46-HAD,1-4.

3-095. View of structure supporting deck, Old Corinth Road Bridge, Sacandaga River, Hadley, Saratoga County, New York, 1885 (rehabilitated 2006). Berlin Iron Bridge Company (East Berlin, CT), builder. Jet T. Lowe, photographer, 1994. P&P,HAER, NY,46-HAD,1-8.

3-096. Elevation details of main and approach trusses, Old Corinth Road Bridge, Sacandaga River, Hadley, Saratoga County, New York, 1885 (rehabilitated 2006). Berlin Iron Bridge Company (East Berlin, CT), builder. Karl Bodensiek, Caroline Schweyer, Luis G. Rosario-Lluveras, delineators, 1994–1995. P&P,HAER,NY,46-HAD,1-,sheet no. 2.

3-094

3-095

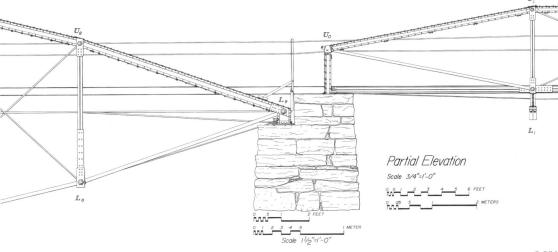

3-096

3-097

3-098

3-099

3-097. Cemetery Road Bridge, Black Creek, Salem Vic., Washington County, New York, 1888. Berlin Iron Bridge Company (East Berlin, CT), builder. Martin Stupich, photographer, 1987. P&P,HAER,NY,58-SAL,1-2.

3-098. Detail of truss panel, Cemetery Road Bridge, Black Creek, Salem Vic., Washington County, New York, 1888. Berlin Iron Bridge Company (East Berlin, CT), builder. Martin Stupich, photographer, 1987. P&P,HAER,NY,58-SAL,1-5.

3-099. Boardman's Bridge, Housatonic River, New Milford, Litchfield County, Connecticut, 1888. Berlin Iron Bridge Company (East Berlin, CT), builder. Jet T. Lowe, photographer, 1994. P&P,HAER, CONN,3-NEMI,1-2.

3-100. Waterville Bridge, Swatara Creek, Appalachian Trail, Green Point Vic., Lebanon County, Pennsylvania, 1890 (originally spanned Little Pine Creek, Waterville; moved 1985). Berlin Iron Bridge Company (East Berlin, CT), builder. Joseph Elliott, photographer, 1997. P&P,HAER,PA,38-GREPO.V,1-2.

3-101. Interior showing Warren stiffening truss, Waterville Bridge, Swatara Creek, Appalachian Trail, Green Point Vic., Lebanon County, Pennsylvania, 1890 (originally spanned Little Pine Creek, Waterville; moved 1985). Berlin Iron Bridge Company (East Berlin, CT), builder. Joseph Elliott, photographer, 1997. P&P,HAER,PA,38-GREPO.V,1-6.

3-102. Pin connection, Waterville Bridge, Swatara Creek, Appalachian Trail, Green Point Vic., Lebanon County, Pennsylvania, 1890 (originally spanned Little Pine Creek, Waterville; moved 1985). Berlin Iron Bridge Company (East Berlin, CT), builder. Joseph Elliott, photographer, 1997. P&P,HAER,PA,38-GREPO.V,1-12.

3-100

3-101

3-102

3-103

3-104

DETAILS

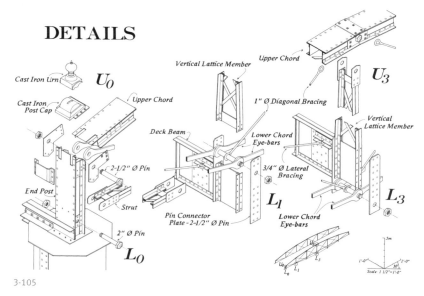

3-105

3-103. Augusta Street Bridge, San Antonio River, San Antonio, Texas, 1890. Berlin Iron Bridge Company (East Berlin, CT), builder. Bruce A. Harms, photographer, 2001. P&P,HAER,TX,15-SANT, 45-2.

South-central Texas is far removed from the Berlin Iron Bridge Company's primary marketing areas in New England and New York State, but an energetic salesman named William Payson convinced officials in San Antonio and nearby counties to purchase its lenticular truss bridges.

3-104. Kelley Crossing Bridge, Plum Creek, Lockhart Vic., Caldwell County, Texas, ca. 1895. Berlin Iron Bridge Company (East Berlin, CT), builder. Joseph Elliott, photographer, ca. 1996. P&P,HAER, TX,28-LOCK.V,2-1.

3-105. Connection details, Kelley Crossing Bridge, Plum Creek, Lockhart Vic., Caldwell County, Texas, ca. 1895. Berlin Iron Bridge Company (East Berlin, CT), builder. Christopher B. Brown, delineator, 1996. P&P,HAER,TX,28-LOCK.V,2-,sheet no. 3.

3-106. Neshanic Station Lenticular Truss Bridge, Raritan River, Neshanic Station, Somerset County, New Jersey, 1896. Berlin Iron Bridge Company (East Berlin, CT), builder. Jet T. Lowe, photographer, 1983. P&P,HAER,NJ,18-NESTA,1-6.

3-107. Interior view, Neshanic Station Lenticular Truss Bridge, Raritan River, Neshanic Station, Somerset County, New Jersey, 1896. Berlin Iron Bridge Company (East Berlin, CT), builder. Jet T. Lowe, photographer, 1983. P&P,HAER,NJ,18-NESTA,1-11.

3-108. Detail of end panels and abutment, Neshanic Station Lenticular Truss Bridge, Raritan River, Neshanic Station, Somerset County, New Jersey, 1896. Berlin Iron Bridge Company (East Berlin, CT), builder. Jet T. Lowe, photographer, 1983. P&P,HAER, NJ,18-NESTA,1-17.

The steel members bracing the portal and the concrete encasing the end posts were added in 1932.

3-106

3-107

3-108

In 1820, the architect Ithiel Town (1784–1844) patented a wooden truss composed of a diagonal lattice of uniform members connecting the top and bottom chords. Town lattice trusses could be built quickly and were used for both railroad and highway spans.

3-109

3-110

3-109. Tucker Toll Bridge, Connecticut River, Bellows Falls, Leavenworth County, Vermont, 1831 (demolished). DETR, photographer, ca. 1900–1910. P&P,LC-D4-70093.

3-110. Wissahickon Railroad Bridge, Wissahickon River, Manayunk Vic., Montgomery County, Pennsylvania, ca. 1835 (demolished). Moncure Robinson, builder. Hand-tinted lithograph, Charles Fenderich, delineator, ca. 1835–1846. P&P,LC-USZ62-1385.

3-111

3-113

3-112

3-114

3-111. Cornish-Windsor Covered Bridge, Connecticut River, Cornish, Sullivan County, New Hampshire, 1866 (rehabilitated 1989). James Tasker and Bela Fletcher, builders. Jet T. Lowe, photographer, 1984. P&P,HAER,NH,10-CORN, 2-3.

This two-span, 460-foot-long structure is the longest covered bridge in the United States.

3-112. Interior, Cornish-Windsor Covered Bridge, Connecticut River, Cornish, Sullivan County, New Hampshire, 1866 (rehabilitated 1989). James Tasker and Bela Fletcher, builders. Jet T. Lowe, photographer, 1984. P&P,HAER, NH,10-CORN,2-10.

3-113. Holliwell Bridge, Middle River, Winterset Vic., Madison County, Iowa, 1879–1880. H. P. Jones and G. K. Foster, builders. Joseph Elliott, photographer, 1995. P&P,HAER,IOWA,61-WINSE.V,1-1.

3-114. Interior, Holliwell Bridge, Middle River, Winterset Vic., Madison County, Iowa, 1879–1880. H. P. Jones and G. K. Foster, builders. Joseph Elliott, photographer, 1995. P&P,HAER,IOWA,61-WINSE.V,1-4.

A wooden arch stiffens the 110-foot Town lattice truss.

3-115

3-116

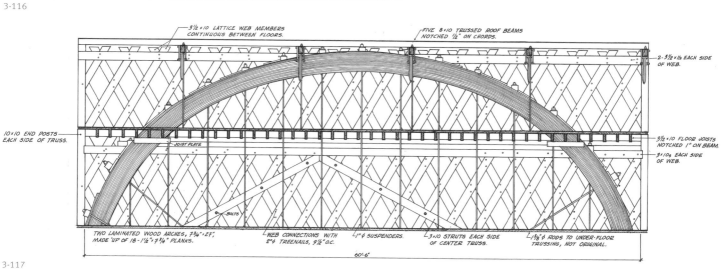

3-117

3-115. E. & T. Fairbanks & Company covered bridge, Sleepers River, St. Johnsbury, Caledonia County, Vermont, ca. 1876 (burned 1972). Lambert Packard, builder. Jack E. Boucher, photographer, 1969. P&P,HAER,VT,3-SAJON,1A-1.

The Fairbanks Company was an innovative and successful manufacturer of scales. During an expansion of the factory, an existing Town lattice-truss road bridge was incorporated into the structure and transformed by the addition of a second story reinforced by a wooden arch.

3-116. Interior showing structure at level of second floor, E. & T. Fairbanks & Company covered bridge, Sleepers River, St. Johnsbury, Caledonia County, Vermont, ca. 1876 (burned 1972). Lambert Packard, builder. Jack E. Boucher, photographer, 1969. P&P,HAER,VT,3-SAJON,1A-8.

3-117. Section, E. & T. Fairbanks & Company covered bridge, Sleepers River, St. Johnsbury, Caledonia County, Vermont, ca. 1876 (burned 1972). Lambert Packard, builder. Eric DeLony, delineator, 1970. P&P,HAER,VT,3-SAJON,1A-,sheet no. 2.

3-118. Upper Bridge at Slate Run, Pine Creek, Slate Run Vic., Lycoming County, Pennsylvania, 1890. Berlin Iron Bridge Company (Berlin, CT). Joseph Elliott, photographer, 1997. P&P,HAER,PA,41-SLARU.V,1-3.

This 202-foot, riveted, wrought-iron through truss is an unusual variation on the lattice type with the diagonal members crossing five times.

3-119. View of portal, Upper Bridge at Slate Run, Pine Creek, Slate Run Vic., Lycoming County, Pennsylvania, 1890. Berlin Iron Bridge Company (Berlin, CT). Joseph Elliott, photographer, 1997. P&P,HAER,PA,41-SLARU.V,1-2.

3-118

3-119

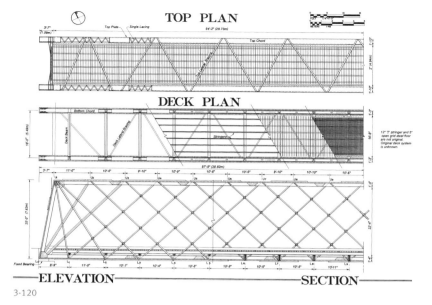

3-120

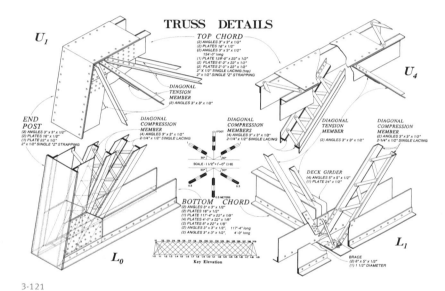

3-121

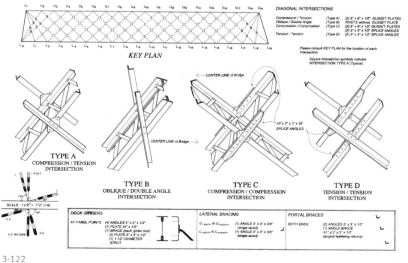

3-122

3-120. Plans and elevation, Upper Bridge at Slate Run, Pine Creek, Slate Run Vic., Lycoming County, Pennsylvania, 1890. Berlin Iron Bridge Company (Berlin, CT). Slavica Bubic and Elizabeth Milnarik, delineators, 1997. P&P,HAER,PA,41-SLARU.V,1-, sheet no. 2.

3-121. Connection details at top and bottom chords, Upper Bridge at Slate Run, Pine Creek, Slate Run Vic., Lycoming County, Pennsylvania, 1890. Berlin Iron Bridge Company (Berlin, CT). Slavica Bubic and Elizabeth Milnarik, delineators, 1997. P&P,HAER,PA,41-SLARU.V,1-,sheet no. 4.

3-122. Truss web intersection details, Upper Bridge at Slate Run, Pine Creek, Slate Run Vic., Lycoming County, Pennsylvania, 1890. Berlin Iron Bridge Company (Berlin, CT). Slavica Bubic and Elizabeth Milnarik, delineators, 1997. P&P,HAER,PA,41-SLARU.V, 1,sheet no. 5.

LONG TRUSS

Stephen H. Long (1784–1864), an officer in the U.S. Army Topographical Service, led mapping expeditions between the Mississippi River and the Rocky Mountains and coined the term "Great American Desert" for the grasslands between them. In the 1830s, he obtained patents for a wooden panel truss system composed of diagonal cross-bracing between vertical posts. This was among the first truss systems proportioned on the basis of mathematical calculations rather than rules of thumb. Also of note was Long's effort to anticipate the behavior of the truss by driving wedges to prestress the diagonal members.

3-123. View of structure during dismantling, Brownsville Covered Bridge, Mill Race Park, Columbus, Bartholomew County, Indiana (originally located over the East Fork of the Whitewater River, Brownsville, IN), 1837–1840 (re-erected with modifications in present location, 1986). Adam Mason, builder. Jack E. Boucher, photographer, 1974. P&P, HAER,IND,81-BROVI,1-6.

3-124. Elevations, Brownsville Covered Bridge, Mill Race Park, Columbus, Bartholomew County, Indiana (originally located over the East Fork of the Whitewater River, Brownsville, IN), 1837–1840 (re-erected with modifications in present location, 1986). Adam Mason, builder. Mark Mattox, delineator, 1971. P&P,HAER,IND,81-BROVI,1-,sheet no. 4.

3-123

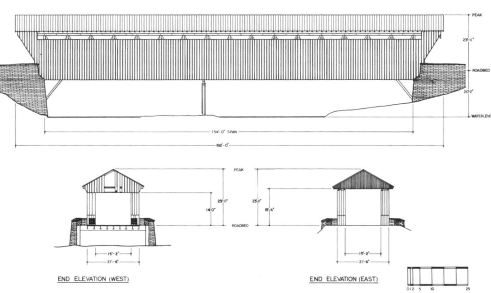

3-124

3-125

3-126. Sections and plans, Blenheim Covered Bridge, Schoharie River, North Blenheim, Schoharie County, New York, 1855. Nichols Montgomery Powers, builder. A. K. Mosley, delineator, 1936. P&P,HABS,NY,48-BLEN,1-,sheet no. 2.

This bridge has a 210-foot clear span and a width of 26 feet, accommodating two lanes of traffic. The Vermont builder, Nichols Montgomery Powers (1817–1887), built the structure with three Long trusses. He reinforced the center truss with an arch.

3-127. Staats Mill Covered Bridge, Tug Fork River, Ripley Vic., Jackson County, West Virginia, 1887. H. T. Hartley, superstructure; Quincy & Grim, masons. West Virginia Department of Transportation, photographer, 1982. P&P,HAER,WVA,18-RIP.V,1-1.

3-128. Interior, Staats Mill Covered Bridge, Tug Fork River, Ripley Vic., Jackson County, West Virginia, 1887. H. T. Hartley, superstructure; Quincy & Grim, masons. West Virginia Department of Transportation, photographer, 1982. P&P,HAER,WVA,18-RIP.V,1-4.

3-125. Blenheim Covered Bridge, Schoharie River, North Blenheim, Schoharie County, New York, 1855. Nichols Montgomery Powers, builder. HABS, photographer, ca. 1936. P&P,HABS,NY,48-BLEN,1-1.

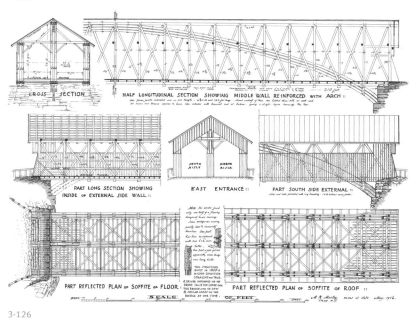

3-126

3-127

3-128

The inventor William Howe (1803–1852) patented a variant on the Long truss in 1840 that substituted wrought-iron rods for the wooden posts defining the panels. Unlike the wooden posts, which were subject to weakening from tensile forces pulling on the connections at the top and bottom chords, the iron rods could be tightened with nuts and turnbuckles (threaded, adjustable coupling). Modified and marketed by his brother-in-law, the architect and railroad builder Amasa Stone (1818–1883), the Howe truss rapidly gained favor with railroad companies and, subsequently, was adapted to all-metal construction.

Composite Wood and Iron Howe Truss

3-129. Jay Covered Bridge, AuSable River, Jay, Essex County, New York, 1857 (covered 1858; shortened, piers added, 1953; rehabilitated 2004). George M. Burt, builder. Martin Stupich, photographer, 1987. P&P,HAER,NY,16-JAY,1-2.

3-130. Interior, Jay Covered Bridge, AuSable River, Jay, Essex County, New York, 1857 (covered 1858; shortened, piers added, 1953; rehabilited 2004). George M. Burt, builder. Martin Stupich, photographer, 1987. P&P,HAER,NY,16-JAY,1-4.

3-129

3-130

3-131

3-133

3-131. Interior, McConnell's Mill Bridge, Slippery Rock Creek, Ellwood City Vic., Lawrence County, Pennsylvania, 1875. J. B. White & Sons, superstructure; Belle & Breckenridge, abutments. Joseph Elliott, photographer, 1997. P&P,HAER,PA,37-ELLCI.V,1-4.

3-132. Truss panel detail, McConnell's Mill Bridge, Slippery Rock Creek, Ellwood City Vic., Lawrence County, Pennsylvania, 1875. J. B. White & Sons, superstructure; Belle & Breckenridge, abutments. Joseph Elliott, photographer, 1997. P&P,HAER,PA,37-ELLCI.V,1-5.

3-133. McConnell's Mill Bridge, Slippery Rock Creek, Ellwood City Vic., Lawrence County, Pennsylvania, 1875. J. B. White & Sons, superstructure; Belle & Breckenridge, abutments. Joseph Elliott, photographer, 1997. P&P,HAER,PA,37-ELLCI.V,1-1.

3-132

3-134

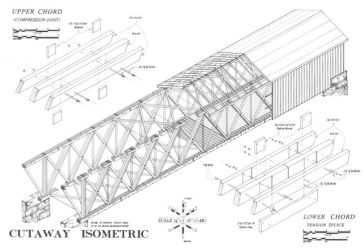

UPPER CHORD
(COMPRESSION JOINT)

CUTAWAY ISOMETRIC

SCALE ¼"=1'-0" (1:48)

LOWER CHORD
TENSION SPLICE

3-135

3-134. Abutment, McConnell's Mill Bridge, Slippery Rock Creek, Ellwood City Vic., Lawrence County, Pennsylvania, 1875. J. B. White & Sons, superstructure; Belle & Breckenridge, abutments. Joseph Elliott, photographer, 1997. P&P,HAER,PA,37-ELLCI.V,1-6.

The steel girders supporting the deck were added around 1959.

3-135. Cutaway isometric view, McConnell's Mill Bridge, Slippery Rock Creek, Ellwood City Vic., Lawrence County, Pennsylvania, 1875. J. B. White & Sons, superstructure; Belle & Breckenridge, abutments. Slavica Bubic and Michael Falser, delineators, 1997. P&P,HAER,PA,37-ELLCI.V,1-,sheet no. 5.

3-136. Truss details, McConnell's Mill Bridge, Slippery Rock Creek, Ellwood City Vic., Lawrence County, Pennsylvania, 1875. J. B. White & Sons, superstructure; Belle & Breckenridge, abutments. Slavica Bubic and Michael Falser, delineators, 1997. P&P,HAER,PA,37-ELLCI.V,1-,sheet no. 6.

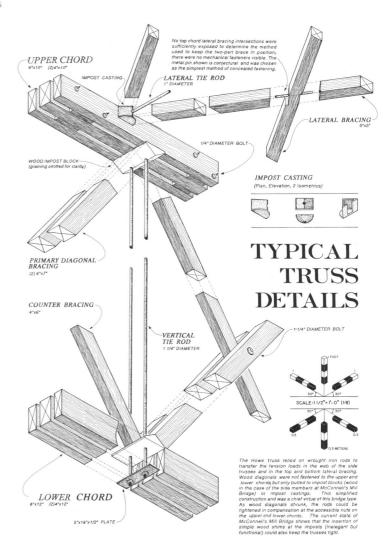

UPPER CHORD
6"x10" (2)4"x10"

No top chord lateral bracing intersections were sufficiently exposed to determine the method used to keep the two-part brace in position; there were no mechanical fasteners visible. The metal pin shown is conjectural and was chosen as the simplest method of concealed fastening.

IMPOST CASTING

LATERAL TIE ROD
1" DIAMETER

LATERAL BRACING
5"x5"

1/4" DIAMETER BOLT

WOOD IMPOST BLOCK
(graining omitted for clarity)

IMPOST CASTING
(Plan, Elevation, 2 Isometrics)

TYPICAL
TRUSS
DETAILS

PRIMARY DIAGONAL
BRACING
(2) 6"x7"

COUNTER BRACING
4"x6"

VERTICAL
TIE ROD
1 1/4" DIAMETER

1-1/4" DIAMETER BOLT

SCALE·1 1/2"=1'-0" (1:8)

LOWER CHORD
6"x12" (2)4"x12"

5"x16"x1/2" PLATE

The Howe truss relied on wrought iron rods to transfer the tension loads in the web of the side trusses and in the top and bottom lateral bracing. Wood diagonals were not fastened to the upper and lower chords, but only butted to impost blocks (wood in the case of the side members at McConnell's Mill Bridge) or impost castings. This simplified construction and was a chief virtue of this bridge type. As wood diagonals shrunk, the rods could be tightened in compensation at the accessible nuts on the upper and lower chords. The current state of McConnell's Mill Bridge shows that the insertion of simple wood shims at the imposts (inelegant but functional) could also keep the trusses tight.

3-136

3-137

3-137. Springdale Covered Bridge, Stranger Creek, Springdale Vic., Leavenworth County, Kansas, ca. 1859 (demolished). John O'Brien, builder. Douglas McCleery, photographer, 1958. P&P,HABS,KANS,52-SPRI.V,1-1.

3-138. Interior, Springdale Covered Bridge, Stranger Creek, Springdale Vic., Leavenworth County, Kansas, ca. 1859 (demolished). John O'Brien, builder. Douglas McCleery, photographer, 1958. P&P,HABS,KANS,52-SPRI.V,1-4.

3-139. Bridgeport Covered Bridge, South Fork of Yuba River, Nevada County, California, 1862. David Ingefield Wood, builder. Jet T. Lowe, photographer, 1984. P&P,HAER,CAL,29-BRIGPO,1-4.

3-140. Interior, Bridgeport Covered Bridge, South Fork of Yuba River, Nevada County, California, 1862. David Ingefield Wood, builder. Jet T. Lowe, photographer, 1984. P&P,HAER,CAL,29-BRIGPO,1-7.

This bridge has a clear span of 208 feet under one Howe truss and 210 feet under the other, and is reinforced by an arch following a design William Howe patented in 1846.

3-138

3-139

3-140

3-141. Detail showing exposed truss, Clark Street Bridge, spanning Boston & Maine Railroad, Belmont, Middlesex County, Massachusetts, 1908. Boston & Maine Railroad, builder. Martin Stupich, photographer, 1990. P&P,HAER,MASS,9-BELM,1-5.

3-142. Clark Street Bridge, spanning Boston & Maine Railroad, Belmont, Middlesex County, Massachusetts, 1908. Boston & Maine Railroad, builder. Martin Stupich, photographer, 1990. P&P,HAER,MASS,9-BELM,1-1.

Boxing a pony truss provides a less expensive covering than building a roofed enclosure.

3-143. View during rehabilitation, Grays River Covered Bridge, Grays River Vic., Wahkiakum County, Washington, 1905 (covered 1908, rehabilitated 1989). Design and construction attributed to Ferguson & Houston (Astoria, OR). Roger Kukas, photographer, 1988. P&P,HAER,WASH,35-GRARI.V,1-4.

3-141

3-142

3-143

3-144

3-145

3-146

3-144. Grays River Covered Bridge, Grays River Vic., Wahkiakum County, Washington, 1905 (covered 1908, rehabilitated 1989). Design and construction attributed to Ferguson & Houston (Astoria, OR). Roger Kukas, photographer, 1991. P&P,HAER,WASH,35-GRARI.V,1-12.

3-145. Harpole Bridge, Palouse River, Great Northern Railway, Colfax Vic., Whitman County, Washington, 1922. Great Northern Railway, builder. Jet T. Lowe, photographer, 1993. P&P,HAER,WASH,38-COLF.V,1-5.

Now converted to vehicular use, this is a rare example of once-common encased wooden Howe railroad bridges.

3-146. Jordan Covered Bridge (Stayton-Jordan Covered Bridge) Pioneer Park, Stayton, Marion County, Oregon, 1937 (moved from Thomas Creek, Scio Vic., Linn County, Oregon, 1987; burned 1994; reconstructed 1998). Oregon State Highway Department, designer; Frank Kaiser, builder. Jerry Robertson, photographer, 1985. P&P,HAER,ORE, 22-SCIO.V,1-7.

The low cost of timber and local building preferences favored the use of covered, wooden Howe trusses for county road bridges in Oregon from the mid-nineteenth through the mid-twentieth centuries. The open-panel design of this bridge shelters the truss while presenting less wind resistance than a fully enclosed structure.

3-147

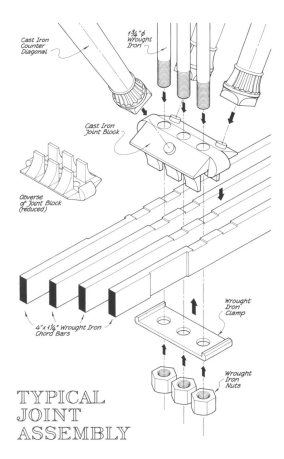

3-148

Iron Howe Truss

3-147. Reading-Halls Station Bridge, Philadelphia & Reading Railroad, near Halls Station, Lycoming County, Pennsylvania, ca. 1846. Richard B. Osborne, engineer. Jet T. Lowe, photographer, 1984. P&P,HAER,PA,41-MUNC.V,1-1.

Originally erected elsewhere as a railroad bridge, this 32-foot (original dimension) Howe pony truss was one of the first all-iron truss bridges erected in the United States. Its designer, Richard B. Osborne (1815–1899), chief engineer for the Philadelphia & Reading Railroad, had built the first such bridge (one truss is in the Smithsonian) about a year before near Manayunk, Pennsylvania.

3-148. Detail of truss, Reading-Halls Station Bridge, Philadelphia & Reading Railroad, near Halls Station, Lycoming County, Pennsylvania, ca. 1846. Richard B. Osborne, engineer. Jet T. Lowe, photographer, 1984. P&P,HAER,PA,41-MUNC.V,1-10.

A remarkable feature of this bridge is the Egyptian Revival–style ornament (such as the lotus motifs) of the cast-iron diagonal members. The vertical rods are wrought iron.

3-149. Connection details, Reading-Halls Station Bridge, Philadelphia & Reading Railroad, near Halls Station, Lycoming County, Pennsylvania, ca. 1846. Richard B. Osborne, engineer. Richard K. Anderson Jr., delineator, 1987. P&P,HAER,PA,41-MUNC.V,1-,sheet no. 5.

3-149

3-150. Oak Knoll Park Bridge, Massillon, Stark County, Ohio, 1859 (moved from original site near Alliance, Ohio, 1899). Joseph Davenport, engineer; C. M. Russell & Company, fabricator. Joseph Elliott, photographer, 1992. P&P,HAER,OHIO,76-MASS,1-2.

For another example of Davenport's use of the Howe truss, see 3-056.

3-151. Oak Knoll Park Bridge, Massillon, Stark County, Ohio, 1859 (moved from original site near Alliance, Ohio, 1899). Joseph Davenport, designer; C. M. Russell & Company, fabricator. Joseph Elliott, photographer, 1992. P&P,HAER,OHIO,76-MASS,1-4.

3-152

3-152. Rush's Mill Bridge, Plum Creek, Sinking Spring Vic., Berks County, Pennsylvania, 1869 (moved from original site spanning Perkiomen Creek, 1979). Simon Dreibelbies, builder. Joseph Elliott, photographer, 1991. P&P,HAER,PA,6-SINSP.V,1-2.

3-153. Sockman Road Bridge, Granny Creek, Fredericktown Vic., Knox County, Ohio, 1873 (demolished). Russell Bridge Company (Massillon, OH), fabricator. Joseph Elliott, photographer, 1992. P&P,HAER,OHIO,42-FRED.V,2-3.

3-154. Connection details, Sockman Road Bridge, Granny Creek, Fredericktown Vic., Knox County, Ohio, 1873 (demolished). Russell Bridge Company (Massillon, OH), fabricator. Attila Kovacs, delineator, 1992. P&P,HAER,OHIO,42-FRED.V,2-,sheet no. 3.

3-153

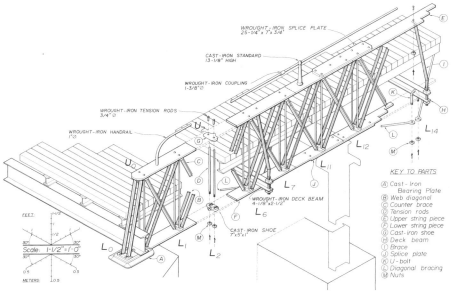

3-154

3-155

Reinforced Concrete Howe Truss

3-155. Broad Street Bridge, Commanche Creek, Mason, Mason County, Texas, 1918. Alamo Construction Company (San Antonio, TX), builder. Joseph Elliott, photographer, 1996. P&P,HAER,TX, 160-MASON,1-6.

This 102-foot bridge is a rare example of a reinforced concrete Howe pony truss.

3-156. View from below of roadway deck and sidewalk, Broad Street Bridge, Commanche Creek, Mason, Mason County, Texas, 1918. Alamo Construction Company (San Antonio, TX), builder. Joseph Elliott, photographer, 1996. P&P,HAER,TX,160-MASON, 1-12.

3-156

In 1844, the engineer Thomas Pratt (1812–1875) and his father, Caleb, patented a design for a wood and iron truss that subsequently proved so well suited for all-metal railroad and highway bridges that it became one of the most commonly used truss types in the nineteenth and early twentieth centuries. Superficially similar to the Howe truss in appearance, the Pratt truss distributes its loads differently, placing the vertical posts in compression and the diagonals in tension. This arrangement reduces the length of the compression members, making them less susceptible to lateral buckling. Many design patents have been issued for variants of its basic form and means of assembly.

Pratt Through Truss

3-157. Walnut Street Bridge, formerly spanning Saucon Creek, Hellertown, Northampton County, Pennsylvania, ca. 1860 (rehabilitated 1999). Charles N. Beckel (Bethlehem, PA), builder. Joseph Elliott, photographer, 1991. P&P,HAER,PA,48-HELLT,3-3.

3-158. Connection details, Walnut Street Bridge, formerly spanning Saucon Creek, Hellertown, Northampton County, Pennsylvania, ca. 1860 (rehabilitated 1999). Charles N. Beckel (Bethlehem, PA), builder. Monica Korsós, delineator, 1991. P&P,HAER,PA,48-HELLT,3-,sheet no. 3.

3-159. Connections at lower chord, Walnut Street Bridge, formerly spanning Saucon Creek, Hellertown, Northampton County, Pennsylvania, ca. 1860 (rehabilitated 1999). Charles N. Beckel (Bethlehem, PA), builder. Joseph Elliott, photographer, 1991. P&P,HAER,PA,48-HELLT,3-9.

This truss, composed of cast- and wrought-iron members, has a span of 55 feet. The wrought-iron rods forming the lower chord, the cast-iron beams supporting the deck, and the cast-iron posts intersect at a connection patented by Francis C. Lowthorp in 1857.

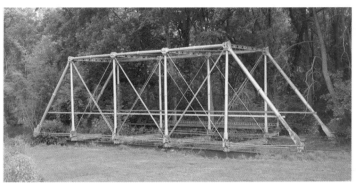

3-157

3-159

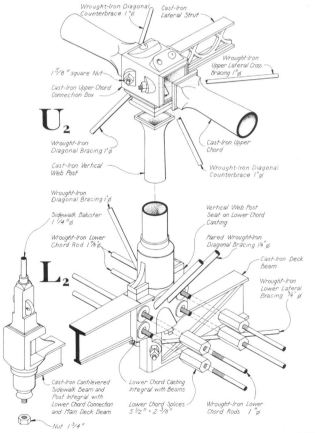

3-158

3-160

3-161

3-162

3-160. Keystone Bridge Company bridge (Fairground Street Bridge), spanning Illinois Central Railroad, Vicksburg, Mississippi, assembled 1895. John Piper and Jacob H. Linville, engineers; Keystone Bridge Company (Pittsburgh, PA), builder. Jet T. Lowe, photographer, 1986. P&P,HAER,MISS,75-VICK,20-1.

This and the White Water Creek Bridge (3-161) exemplify how bridges were recycled in the nineteenth and early twentieth centuries as rail lines were upgraded. The White Water Creek Bridge and one span of the Vicksburg bridge were fabricated in 1872 as components of a series of seven Pratt through trusses added to the western approach of the railroad bridge at Dubuque, Iowa, erected in 1868. Reconfiguration of the approach in 1887 rendered the spans superfluous, and they were sold for use as road bridges. For the bridge over its tracks in Vicksburg, the Illinois Central attached one of the Dubuque spans (purchased in 1894 and moved to Vicksburg a year later) to another Keystone span of similar vintage acquired from an unknown location.

3-161. White Water Creek Bridge, Bernard Vic., Dubuque County, Iowa (formerly the approach span of the Dubuque-Dunleith Railroad Bridge, Mississippi River), 1872. John Piper and Jacob H. Linville, engineers; Keystone Bridge Company (Pittsburgh, PA), builder. Joseph Elliott, photographer, 1995. P&P,HAER,IOWA,31-BERN.V,1-2.

Like the Keystone Bridge Company Bridge in Vicksburg (3-160), this span originally formed part of the western approach of the railroad bridge at Dubuque. It presumably was purchased by a contractor in 1889 who then sold it to Dubuque County for use as a wagon bridge.

3-162. Interior view at mid span, White Water Creek Bridge, Bernard Vic., Dubuque County, Iowa (formerly the approach span of the Dubuque-Dunleith Railroad Bridge, Mississippi River), 1872. John Piper and Jacob H. Linville, engineers; Keystone Bridge Company (Pittsburgh, PA), builder. Joseph Elliott, photographer, 1995. P&P,HAER,IOWA,31-BERN.V,1-4.

3-163

3-164

3-163. Connections at deck, White Water Creek Bridge, Bernard Vic., Dubuque County, Iowa (formerly the approach span of the Dubuque-Dunleith Railroad Bridge, Mississippi River), 1872. John Pipe and Jacob H. Linville, engineers; Keystone Bridge Company (Pittsburgh, PA), builder. Joseph Elliott, photographer, 1995. P&P,HAER,IOWA,31-BERN.V,1-8.

3-164. Top chord and column connection, White Water Creek Bridge, Bernard Vic., Dubuque County, Iowa (formerly the approach span of the Dubuque-Dunleith Railroad Bridge, Mississippi River), 1872. John Piper and Jacob H. Linville, engineers; Keystone Bridge Company (Pittsburgh, PA), builder. Joseph Elliott, photographer, 1995. P&P,HAER,IOWA,31-BERN.V,1-6.

Linville's patented, wrought-iron built-up column was the Keystone Company's counterpart to the Phoenix column (3-238–3-239) patented by its competitor, the Phoenix Iron Company.

3-165. Connection details, White Water Creek Bridge, Bernard Vic., Dubuque County, Iowa (formerly the approach span of the Dubuque-Dunleith Railroad Bridge, Mississippi River), 1872. John Pipe and Jacob H. Linville, engineers; Keystone Bridge Company (Pittsburgh, PA), builder. Caroline Schweyer, delineator, 1995. P&P,HAER,IOWA,31-BERN.V,1-,sheet no. 4.

Like most truss bridges erected in the second half of the nineteenth century, this one has pinned connections that allow members to respond to strains by revolving around the joint. The arrangement enables designers to analyze the truss as a statically determinate structure using simple algebraic formulas.

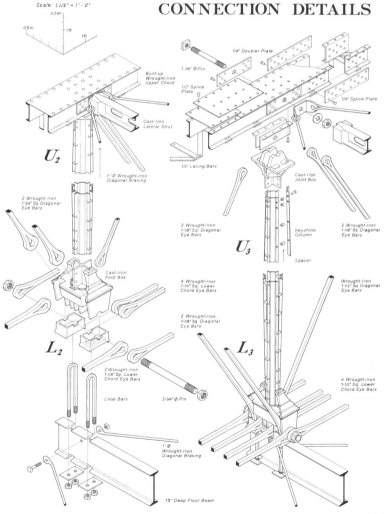

CONNECTION DETAILS

3-165

3-166

3-167. Pine Mill Bridge, Pine Creek, Muscatine Vic., Muscatine County, Iowa, 1878. Wrought Iron Bridge Company (Canton, OH), builder. Joseph Elliott, photographer, 1995. P&P,HAER,IOWA,70-MUSCA.V,2-3.

3-168. Pecatonica River Bridge, Freeport Vic., Stephenson County, Illinois, 1879. Wrought Iron Bridge Company (Canton, OH), builder. R. J. Ware, photographer, 1982. P&P,HAER,ILL,89-FREP.V,1-3.

The Wrought Iron Brige Company marketed its products and construction services throughout the Midwest. These Pratt through trusses are examples of one of the company's most popular designs.

3-169. Gospel Street Bridge, Lick Creek, Paoli, Orange County, Indiana, 1880. Cleveland Bridge & Iron Company (Cleveland, OH), builder. Jack E. Boucher, photographer, 1974. P&P,HAER,IND,59-PAOL,2-5.

3-167

3-168

3-169

3-170

3-171

3-170. Bridge No. 6023, spanning Norfolk-Southern railroad tracks, Nokesville Vic., Prince William County, Virginia, 1882. Virginia Midland Railway and Keystone Bridge Company, designers; Keystone Bridge Company (Pittsburgh, PA), builder. James DuSel, photographer, 1994. P&P,HAER, VA,77-NOKV.V,1-5.

Originally built as a railroad bridge, this 74-foot, pin-connected Pratt through truss was moved to its present location for use as a highway bridge after 1905.

3-171. Connections at portal, Bridge No. 6023, spanning Norfolk-Southern railroad tracks, Nokesville Vic., Prince William County, Virginia, 1882. Virginia Midland Railway and Keystone Bridge Company, designers; Keystone Bridge Company (Pittsburgh, PA), builder. James DuSel, photographer, 1994. P&P,HAER,VA,77-NOKV.V,1-15.

3-172. Deck structure, Bridge No. 6023, spanning Norfolk-Southern railroad tracks, Nokesville Vic., Prince William County, Virginia, 1882. Virginia Midland Railway and Keystone Bridge Company, designers; Keystone Bridge Company (Pittsburgh, PA), builder. James DuSel, photographer, 1994. P&P,HAER,VA, 77-NOKV.V,1-17.

3-173. Bearing at abutment, Bridge No. 6023, spanning Norfolk-Southern railroad tracks, Nokesville Vic., Prince William County, Virginia, 1882. Virginia Midland Railway and Keystone Bridge Company, designers; Keystone Bridge Company (Pittsburgh, PA), builder. James DuSel, photographer, 1994. P&P,HAER,VA,77-NOKV.V,1-21.

3-172

3-173

3-174

3-175

3-176

3-174. Cartersville Bridge, James River, Cartersville Vic., Cumberland County, Virginia, 1884 (all but two spans demolished 1972). Cartersville Bridge Company, builder. Richard J. Pollak, photographer, 1970. P&P,HAER,VA,25-CART.V,1-1.

This pin-connected, Pratt through-truss highway bridge is a late instance of a composite structure employing wood for the compression members (end posts, upper chord, vertical posts), wrought iron for the tension members (diagonals, bottom chord), and cast iron (connectors).

3-175. Provo River Bridge, Denver & Rio Grande Western Railroad, Orem Vic., Orem County, Utah, 1884 (originally spanned Green River; moved to Price River 1901; moved to present location 1919). Union Bridge Company (Athens, PA), builder. Jack E. Boucher, photographer, 1971. P&P,HAER,UTAH,25-OLMS,1-1.

3-176. Nineteenth Street Bridge, South Platte River, Denver, Colorado, 1887. Missouri Valley Bridge & Iron Company (Leavenworth, KS), builder. Troy Ostendorf, photographer, 1991. P&P,HAER,COLO,16-DENV,58-4.

3-177. Selby Avenue Bridge, spanning Short Line Railways track, St. Paul, Minnesota, 1890 (demolished). Andreas W. Munster, engineer; Edge Moor Bridge Works (Wilmington, DE), prime contractor. Fredric L. Quivik, photographer, 1992. P&P,HAER,MINN, 62-SAIPA,34-4.

3-178. Bearing of skewed truss at intermediate pier, Selby Avenue Bridge, spanning Short Line Railways track, St. Paul, Minnesota, 1890 (demolished). Andreas W. Munster, engineer; Edge Moor Bridge Works (Wilmington, DE), prime contractor. Fredric L. Quivik, photographer, 1992. P&P,HAER,MINN,62-SAIPA,34-17.

3-179. Wisconsin & Michigan Railroad bridge, Menominee River, Marinette County, Wisconsin, and Menominee County, Michigan, 1894 (converted to highway use 1938; demolished). F. S. Brown & Company (Chicago, IL), contractor. Dietrich Floeter, photographer, 1989. P&P,HAER,WIS,38-WAG,1-2.

3-180

3-181

3-182

3-183

3-180. Knaggs Bridge, Shiawassee River, Bancroft Vic., Shiawassee County, Michigan, 1895 (demolished ca. 1988). King Bridge Company (Cleveland, OH), builder. Dietrich Floeter, photographer, 1987. P&P,HAER,MICH,78-BANC.V,1-3.

3-181. Four Mile Bridge, Elk River, Steamboat Springs Vic., Routt County, Colorado, 1900. Wrought Iron Bridge Company (Canton, OH). Jim Yannaccone, photographer, 1988. P&P,HAER,COLO,54-STESP.V,11.

3-182. Joshua Falls Railroad Bridge, Chesapeake & Ohio Railway (now CSX Line), James River, Lynchburg Vic., Campbell County, Virginia, 1901. Pencoyd Iron Works (Pencoyd, PA) builder. Eugene B. Barfield, photographer, 1993. P&P,HAER,VA,16-LYMBU.V,1-2.

3-183. Green River Bridge, Daniel Vic., Sublette County, Wyoming, ca. 1905. Western Bridge and Construction Company (Omaha, NE), builder. Clayton B. Fraser, photographer, 1982. P&P,HAER, WYO,18-DAN.V,1-1.

3-184. McClure Bridge, north fork of Palouse River, Palouse Vic., Whitman County, Washington, 1908 (demolished). E. G. Murray, county engineer, designer; O. H. Horton, builder. Harvey S. Rice, photographer, 1989. P&P,HAER,WASH,38-PALO.V,1-4.

Erected at a time when most Pratt trusses were all steel, this 120-foot through truss has compression members of wood, a material readily available in Washington.

3-185. Detail of truss panel, McClure Bridge, north fork of Palouse River, Palouse Vic., Whitman County, Washington, 1908 (demolished). E. G. Murray, county engineer, designer; O. H. Horton, builder. Harvey S. Rice, photographer, 1989. P&P,HAER,WASH,38-PALO.V,1-7.

The top surface of the wooden upper chord has metal flashing to protect it from rain.

3-186. Detail at abutment showing iron bearing shoe and lower chord, McClure Bridge, north fork of Palouse River, Palouse Vic., Whitman County, Washington, 1908 (demolished). E. G. Murray, county engineer, designer; O. H. Horton, builder. Harvey S. Rice, photographer, 1989. P&P,HAER,WASH, 38-PALO.V,1-5.

3-187

3-187. Big Island Bridge, Green River, Green River Vic., Sweetwater County, Wyoming, 1909–1910. Charles G. Sheely, builder. Clayton B. Fraser, photographer, 1982. P&P,HAER,WYO,19-GRERI.V,1-3.

3-188. Bert Parsons Bridge, Fisher River, Cornell Vic., Chippewa County, Wisconsin, 1914. Wisconsin State Highway Commission, designer. Jerry Mathiason, photographer, 1994. P&P,HAER,WIS,9-CORN.V,1-1.

This 90-foot, steel Pratt through truss is an example of the Wisconsin State Highway Commission's use of standardized designs, which began in 1911. The rigid bolted and riveted connections reflect the Commission's gradual abandonment of the formerly ubiquitous pinned connections on the principle that rigid-connected trusses were more durable.

3-189. Detail of connections at deck, Bert Parsons Bridge, Fisher River, Cornell Vic., Chippewa County, Wisconsin, 1914. Wisconsin State Highway Commission, designer. Jerry Mathiason, photographer, 1994. P&P,HAER,WIS,9-CORN.V,1-8.

3-188

3-189

3-190. St. Mary Bridge and Siphon, St. Mary River, Sabb Vic., Glacier County, Montana, 1915. Minneapolis Bridge Company, builder. Jet T. Lowe, photographer, 1980. P&P,HAER,MONT,18-BABB.V,1-4.

This vehicular bridge and aqueduct is part of a 29-mile irrigation canal carrying water across the divide separating the St. Mary River from the Milk River watershed.

3-191. Cypress Creek Bridge, Perry Vic., Perry County, Arkansas, ca. 1915 (demolished 1993). Jeff Holder and Michael Swanda, photographers, 1988. P&P,HAER,ARK,53-PER.V,1-4.

3-192. Interior, Cypress Creek Bridge, Perry Vic., Perry County, Arkansas, ca. 1915 (demolished 1993). Jeff Holder and Michael Swanda, photographers, 1988. P&P,HAER,ARK,53-PER.V,1-5.

Accommodating site conditions, this 80-foot, rigid-connected span is skewed, requiring the western truss (at right) to be offset by one panel.

3-193

3-194

3-193. Enloe Bridge No. 90021, Red River, Wolverton Vic., Wilkin County, Minnesota, 1917. Great Northern Bridge Company (Minneapolis, MN), builder. Mike Whyte, photographer, 1993. P&P,HAER,MINN,84-WOLV.V, 1-3.

3-194. Blackwell Bridge, Beaverdam Creek, Heardmont Vic., Elbert County, Georgia, 1917. Austin Brothers Bridge Company (Atlanta, GA), builder. Dennis O'Kain, photographer. P&P,HAER,GA,53-HEAR.V,1-8.

3-195

Pratt Pony Truss

3-195. New Hampton Bridge (Musconetcong Bridge), Musconetcong River, New Hampton Vic., Hunterdon County, New Jersey, 1868. William and Charles Cowin, Cowin Iron Works (Lambertville, NJ), designers and builders. Joseph Elliott, photographer, 1991. P&P,HAER,NJ,10-HAMP.V,1-5.

3-196. Detail of connections at deck, New Hampton Bridge (Musconetcong Bridge), Musconetcong River, New Hampton Vic., Hunterdon County, New Jersey, 1868. William and Charles Cowin, Cowin Iron Works (Lambertville, NJ), designers and builders. Joseph Elliott, photographer, 1991. P&P,HAER,NJ,10-HAMP.V,1-11.

3-197. Detail of cast-iron bearing and stone masonry abutment, New Hampton Bridge (Musconetcong Bridge), Musconetcong River, New Hampton Vic., Hunterdon County, New Jersey, 1868. William and Charles Cowin, Cowin Iron Works (Lambertville, NJ), designers and builders. Joseph Elliott, photographer, 1991. P&P,HAER,NJ,10-HAMP.V,1-7.

3-196

3-197

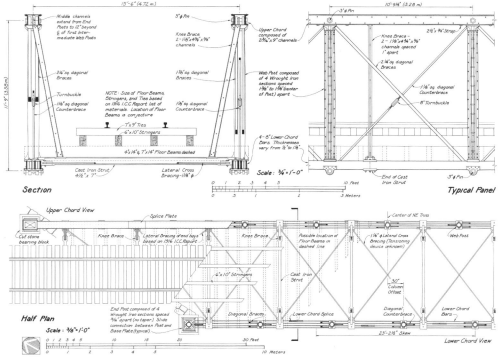

3-198. Stewartstown Railroad Bridge, Northern Central Line (now Stewartstown Railroad), spanning Valley Road, Stewartstown Vic., York County, Pennsylvania, 1870 (moved to present site from Jones Falls, Baltimore, 1885). Jacob Hays Linville, engineer; Keystone Bridge Company (Pittsburgh, PA), fabricator. Joseph Elliott, photographer, 1991. P&P,HAER,PA,67-STEW.V,1-3.

3-199. Section, elevation, and plan, Stewartstown Railroad Bridge, Northern Central Line (now Stewartstown Railroad), spanning Valley Road, Stewartstown Vic., York County, Pennsylvania, 1870 (moved to present site from Jones Falls, Baltimore, 1885). Jacob Hays Linville, engineer; Keystone Bridge Company (Pittsburgh, PA), fabricator. Christine Ussler, delineator, 1991. P&P,HAER,PA,67-STEW.V,1-,sheet no. 2.

3-200. Old Mill Road Bridge, Saucon Creek, Hellertown Vic., Northampton County, Pennsylvania, 1870. Charles Nathaniel Beckel, Beckel Iron Foundry & Machine Shop (Bethlehem, PA), builder. Jet T. Lowe, photographer, 1984. P&P,HAER,PA,48-HELLT,2-10.

3-201. Detail of truss panel, Old Mill Road Bridge, Saucon Creek, Hellertown Vic., Northampton County, Pennsylvania, 1870. Charles Nathaniel Beckel, Beckel Iron Foundry & Machine Shop (Bethlehem, PA), builder. Jet T. Lowe, photographer, 1984. P&P,HAER,PA,48-HELLT,2-7.

3-202. Connection details, Old Mill Road Bridge, Saucon Creek, Hellertown Vic., Northampton County, Pennsylvania, 1870. Charles Nathaniel Beckel, Beckel Iron Foundry & Machine Shop (Bethlehem, PA), builder. Coy Burnet and Monika Korsós, delineators, 1986–1991. P&P,HAER,PA,48-HELLT,2-,sheet no. 3.

The design of the integral cast-iron column and beam is based on patents held by Francis Lowthorp.

3-200

3-201

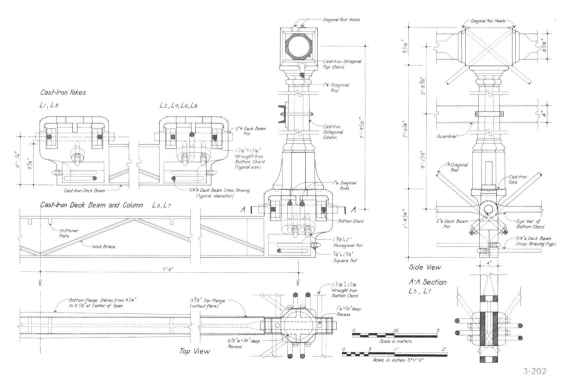

3-202

3-203

3-204

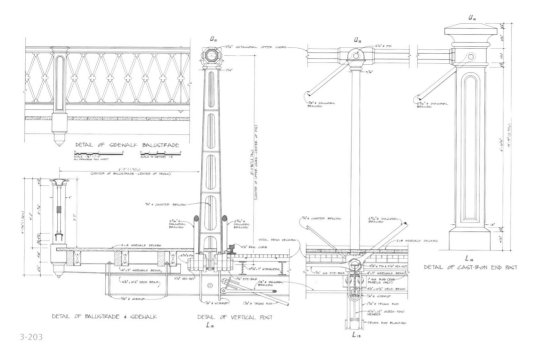

3-205

3-203. Details, West Main Street Bridge, Raritan River, Clinton, Hunterdon County, New Jersey, 1870. Francis C. Lowthorp, engineer; Cowin Iron Works (Lambertville, NJ), builder. Carolyn Givens, delineator, 1985. P&P,HAER,NJ,10-CLIN,1-,sheet no. 3.

3-204. West Main Street Bridge, Raritan River, Clinton, Hunterdon County, New Jersey, 1870. Francis C. Lowthorp, engineer; Cowin Iron Works (Lambertville, NJ), builder. Jack E. Boucher, photographer, 1971. P&P,HAER,NJ,10-CLIN,1-5.

3-205. Butler Bridge, Housatonic River, Stockbridge Vic., Berkshire County, Massachusetts, 1881–1882. George S. Morison, engineer; Morison, Field & Company (New York, NY), ironwork; J. H. Burghardt & Company (Curtisville, MA), masonry. Martin Stupich, photographer, 1990. P&P,HAER, MASS,2-STOCK,6-3.

This configuration of the deck at mid-height within the truss often is described as a half-pony truss.

3-206. Scarlets Mill Bridge, spanning former Reading Railroad line, Scarlets Mill, Berks County, Pennsylvania, 1881. John L. Foreman, engineer; Philadelphia & Reading Railroad, builder; Phoenix Iron Company (Phoenixville, PA), manufacturer, iron beams. Joseph Elliott, photographer, 1991. P&P,HAER,PA,6-SCAMI,1-1.

This road bridge is an unusual example of a Pratt truss with a curved upper chord. It also represents a relatively late use of cast iron.

3-207. Detail of cast-iron knee brace, railing post, and bearing block, Scarlets Mill Bridge, spanning former Reading Railroad line, Scarlets Mill, Berks County, Pennsylvania, 1881. John L. Foreman, engineer; Philadelphia & Reading Railroad, builder; Phoenix Iron Company (Phoenixville, PA), manufacturer, iron beams. Joseph Elliott, photographer, 1991. P&P,HAER,PA,6-SCAMI,1-10.

3-208. Fosnaugh Bridge, Scippo Creek, Stoutsville Vic., Fairfield County, Ohio, 1891 (demolished). Hocking Valley Bridge Works (Lancaster, OH), builder. Louise T. Cawood, photographer, 1986. P&P,HAER,OHIO,23-STOVI.V,1-1.

3-209

3-210

3-211

3-209. Wells County Bridge No. 74, Rock Creek, Bluffton Vic., Wells County, Indiana, 1903. Indiana Bridge Company (Muncie, IN), builder. Camille B. Fife and Thomas W. Salmon II, photographers, 1997. P&P,HAER,IND,90-BLUFF.V, 1-2.

3-210. Wallace Mill Bridge, Little Calfpasture River, Craigsville, Augusta County, Virginia, 1914 (dismantled). Champion Bridge Company (Wilmington, OH), builder. Bruce A. Harms, photographer, 1994. P&P,HAER,VA,8-CRAIGV.V,1-3.

The arrangement of the truss with vertical end-post legs extending below the deck into the abutment (concrete, in this instance) recalls the legs of a bed frame and is sometimes referred to as a bedstead truss. This bridge is to be re-erected at the Museum of Frontier Culture in Staunton, Virginia.

3-211. End posts and abutment, Wallace Mill Bridge, Little Calfpasture River, Craigsville, Augusta County, Virginia, 1914. Champion Bridge Company (Wilmington, OH), builder. Bruce A. Harms, photographer, 1994. P&P,HAER,VA,8-CRAIGV.V,1-9.

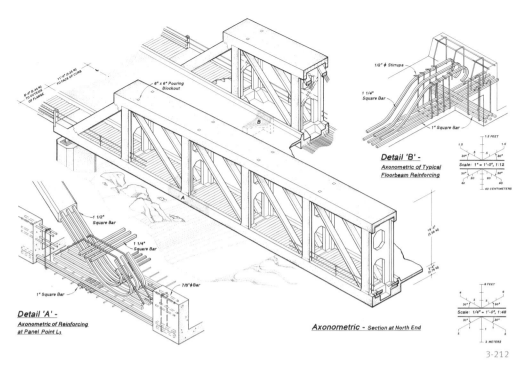

Detail 'B' -
Axonometric of Typical
Floorbeam Reinforcing

Detail 'A' -
Axonometric of Reinforcing
at Panel Point L₃

Axonometric - Section at North End

3-212

3-212. Axonometric view and details, McMillin Bridge, Puyallup River, McMillin, Pierce County, Washington, 1934. Homer M. Hadley, Portland Cement Association, and George Rundman, W. H. Witt Company, engineers; Dolph Jones, builder. Erin M. Doherty, delineator, 1993. P&P,HAER,WASH,27-MCMIL,1-,sheet no. 3.

3-213. McMillin Bridge, Puyallup River, McMillin, Pierce County, Washington, 1934. Homer M. Hadley, Portland Cement Association, and George Rundman, W. H. Witt Company, engineers; Dolph Jones, builder. Jet T. Lowe, photographer, 1993. P&P,HAER,WASH,27-MCMIL,1-2.

This 170-foot span is a rare example of a Pratt truss constructed of reinforced concrete. The trusses have a depth of 20 feet at the center of the span. Lateral reinforcing is contained within the 7-foot width.

3-214. McMillin Bridge, Puyallup River, McMillin, Pierce County, Washington, 1934. Homer M. Hadley, Portland Cement Association, and George Rundman, W. H. Witt Company, engineers; Dolph Jones, builder. Jet T. Lowe, photographer, 1993. P&P,HAER,WASH,27-MCMIL,1-5.

The sidewalks pass through 7-foot-high openings within the interior of the trusses.

3-213

3-214

3-215

Pratt Deck Truss

3-215. Dearborn River High Bridge, Augusta Vic., Lewis and Clark County, Montana, 1897 (rehabilitated 2003). King Bridge Company (Cleveland, OH), builder. Jet T. Lowe, photographer, 1980. P&P,HAER,MONT,25-AUG.V,1-4.

The roadway of this 160-foot span is attached to the vertical posts of the side trusses slightly above their midpoints, forming an unusual configuration that can be described as a half-deck or half-pony truss.

3-216. Main Channel Bridge, spanning Clark Fork, Thompson Falls, Sanders County, Montana, 1911. William Pierce Cowles (Minneapolis, MN), engineer; O. E. Peppard (Missoula, MT), builder. Jet T. Lowe, photographer, 1980. P&P,HAER,MONT,45-THOFA,2-8.

3-217. Powder River Bridge, Arvada Vic., Sheridan County, Wyoming, 1932–1933. W. P. Roscoe Company, builder (Billings, MT). Clayton B. Fraser, photographer, 1992. P&P,HAER,WYO,17-ARVA.V,1-2.

This continuous riveted rigid-connected Pratt deck truss illustrates how steel bridge construction of the 1930s differs from that of the first decades of the century, as seen in the Main Channel Bridge (3-216), which is a pin-connected simple span.

3-216

3-217

PARKER TRUSS

Charles H. Parker (ca. 1842–1897) obtained a series of patents between 1868 and 1871 for a variation of the Pratt through truss known as the Parker truss and manufactured it at the National Bridge & Iron Works (1868–1875) in Boston. In its early form, seen in the Elm Street (3-218) and North Village (3-219) bridges, the upper chord is shaped like an arch with inclined end panels. The configuration requires less material than the standard Pratt truss because it allows the truss at the center of the span (where bending is greatest) to be deeper than at the ends. Bridges with trusses composed of five upper chord segments are commonly called camelbacks.

3-218. Elm Street Bridge, Ottauquechee River, Woodstock, Windsor County, Vermont, 1870 (reinforced with Warren truss below the deck ca. 1910; rehabilitated). Charles H. Parker, engineer; National Bridge & Iron Works (Boston, MA) builder. Dennis Zembala, photographer, 1976. P&P,HAER, VT,14-WOOD,8-16.

3-219. North Village Bridge, French River, Webster, Worcester County, Massachusetts, 1871. Charles H. Parker, engineer; National Bridge & Iron Works (Boston, MA), builder. Martin Stupich, photographer, 1990. P&P,HAER,MASS,14-WEB,2-11.

3-218

3-219

3-220

3-221. Manchester Street Bridge, Baraboo River, Baraboo, Sauk County, Wisconsin, 1884. Milwaukee Bridge & Iron Works, builder (Milwaukee, WI). John N. Vogel, photographer, ca. 1986. P&P,HAER,WIS,56-BARAB,3-1.

This is a classic example of a camelback truss.

3-221. Rockport Bridge, Ouachita River, Malvern, Hot Spring County, Arkansas, 1900 (demolished 1990). O. W. Childs (St. Louis, MO), engineer; Stupp Brothers Bridge & Iron Company (St. Louis, MO), builder. Louise T. Cawood, photographer, 1988. P&P,HAER,ARK,30-MALV,1-2.

3-222. Attica Bridge, Deer Creek, Logansport Vic., Cass County, Indiana, 1904. Attica Bridge Company (Attica, IN), builder. Steve Kovack, photographer, 1989. P&P,HAER,IND,9-YOAM.V,1-7.

3-223. Detail of riveted connection, Attica Bridge, Deer Creek, Logansport Vic., Cass County, Indiana, 1904. Attica Bridge Company (Attica, IN), builder. Steve Kovack, photographer, 1989. P&P,HAER,IND,9-YOAM.V,1-11.

3-221

3-222

3-223

3-224. Kinsey Bridge, Chicago, Milwaukee, St. Paul & Pacific Railroad, Yellowstone River, Miles City Vic., Custer County, Montana, ca. 1907. Jet T. Lowe, photographer, 1980. P&P,HAER,MONT,9-MILCI.V,2-3.

3-225. White's Ferry Bridge, Spoon River, Blyton Vic., Fulton County, Illinois, 1910. Springfield Bridge & Iron Company (Springfield, IL), builder. Roger McCredie, photographer, 1992. P&P,HAER, ILL,29-BLY.V,1-3.

3-226. Park Avenue Bridge, San Francisco River, Clifton, Greenlee County, Arizona, 1917–1918. Midland Bridge Company (Kansas City, MO), builder. Robert G. Graham, photographer, 1993. P&P,HABS,ARIZ,6-CLIFT,11-1.

3-224

3-225

3-226

3-227

3-228

3-229

3-227. Enterprise Bridge, Smoky Hill River, Enterprise, Dickinson County, Kansas, 1924–1925 (demolished 1988). Yancey Construction Company (Abilene, KS), builder. Larry Kasbulas, photographer, 1985. P&P,HAER,KANS,21-ENPRI,1-20.

3-228. Upper chord riveted connections, Enterprise Bridge, Smoky Hill River, Enterprise, Dickinson County, Kansas, 1924–1925 (demolished 1988). Yancey Construction Company (Abilene, KS), builder. Larry Kasbulas, photographer, 1985. P&P,HAER,KANS,21-ENPRI,1-6.

3-229. Rocker bearing, Enterprise Bridge, Smoky Hill River, Enterprise, Dickinson County, Kansas, 1924–1925 (demolished 1988). Yancey Construction Company (Abilene, KS), builder. Larry Kasbulas, photographer, 1985. P&P,HAER,KANS,21-ENPRI, 1-12.

3-230

3-230. Gaylordsville Bridge, Housatonic River, New Milford, Litchfield County, Connecticut, 1926. Berlin Construction Company (Berlin, CT), builder. Robert Moore, photographer, 1987. P&P,HAER,CONN,3-NEMI,3-2.

3-231. Sanders Ferry Bridge, Savannah River, Anderson County, South Carolina, 1927. Austin Bridge Company (Atlanta, GA), builder. Dennis O'Kain, photographer, ca. 1980. P&P,HAER,SC,4-SAVRI,1-5.

3-231

3-232

3-233

3-234

3-232. Jensen Bridge, Green River, Jensen, Uintah County, Utah, 1933 (demolished 1994). Maurice Housecroft, Utah State Road Commission, designer; James J. Burke & Company (Salt Lake City, UT), builder. John Pendlebury, photographer, 1987. P&P,HAER,UTAH,24-JENS,1-3.

3-233. CTH D Bridge, Eau Galle River, Eau Galle Vic., Dunn County, Wisconsin, 1934–1935. State Highway Commission, designer; Worden-Allen Company (Milwaukee, WI), builder. John N. Vogel, photographer, 1996. P&P,HAER,WIS, 17-EAGA.V,1-2.

From the 1910s to the 1930s, the Wisconsin State Highway Commission specified Parker and Pratt trusses in its standardized designs for steel through trusses.

3-234. Detail of truss panels, CTH D Bridge, Eau Galle River, Eau Galle Vic., Dunn County, Wisconsin, 1934–1935. State Highway Commission, designer; Worden-Allen Company (Milwaukee, WI), builder. John N. Vogel, photographer, 1996. P&P,HAER,WIS,17-EAGA.V,1-7.

In the late nineteenth century, bridge engineers developed a number of variations on the Pratt truss that increased strength and rigidity by adding braces within the truss panels.

3-235

Baltimore Truss

Introduced by the Baltimore & Ohio Railroad in 1871, this subdivided truss is characterized by diagonal braces and end posts.

3-235. Walnut Street Bridge, Susquehanna River, Harrisburg, Pennsylvania, 1889–1890. Albert Lucius (New York, NY), consulting engineer; Dean & Westbrook Bridge Company (New York, NY), contractor; Phoenix Bridge Company (Phoenixville, PA), fabricator. Rob Tucher, photographer, 1996. P&P,HAER,PA,22-HARBU,25-6.

This bridge consists of twelve 175-foot and three 240-foot pin-connected, wrought-iron Baltimore trusses. Three spans were lifted off their piers by ice floes in 1996.

3-236. Bird's-eye view and map, Walnut Street Bridge, Susquehanna River, Harrisburg, Pennsylvania, 1889–1890. Albert Lucius (New York, NY), consulting engineer; Dean & Westbrook Bridge Company (New York, NY), contractor; Phoenix Bridge Company (Phoenixville, PA), fabricator. Lou A. McCrory and John R. Bowie, delineators, 1996. P&P,HAER,PA,22-HARBU,25-,sheet no. 1.

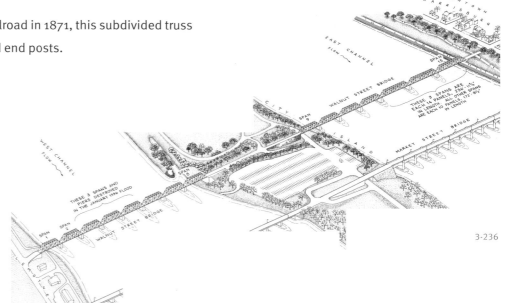

3-236

3-237

3-237. Interior view, Walnut Street Bridge, Susquehanna River, Harrisburg, Pennsylvania, 1889–1890. Albert Lucius (New York, NY), consulting engineer; Dean & Westbrook Bridge Company (New York, NY), contractor; Phoenix Bridge Company (Phoenixville, PA), fabricator. Rob Tucher, photographer, 1996. P&P,HAER,PA,22-HARBU,25-26.

3-238. Truss assembly, Walnut Street Bridge, Susquehanna River, Harrisburg, Pennsylvania, 1889–1890. Albert Lucius (New York, NY), consulting engineer; Dean & Westbrook Bridge Company (New York, NY), contractor; Phoenix Bridge Company (Phoenixville, PA), fabricator. Lou A. McCrory and John R. Bowie, delineators, 1996. P&P,HAER,PA,22-HARBU,25-,sheet no. 4.

The posts, upper chords, and struts are composed of the company's patented (1862) Phoenix columns.

3-239. Connection details, Walnut Street Bridge, Susquehanna River, Harrisburg, Pennsylvania, 1889–1890. Albert Lucius (New York, NY), consulting engineer; Dean & Westbrook Bridge Company (New York, NY), contractor; Phoenix Bridge Company (Phoenixville, PA), fabricator. Lou A. McCrory and John R. Bowie, delineators, 1996. P&P,HAER,PA,22-HARBU,25-,sheet no. 5.

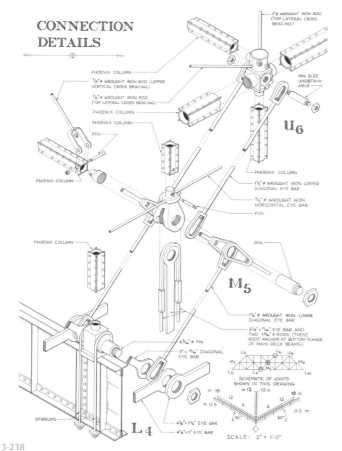

3-238

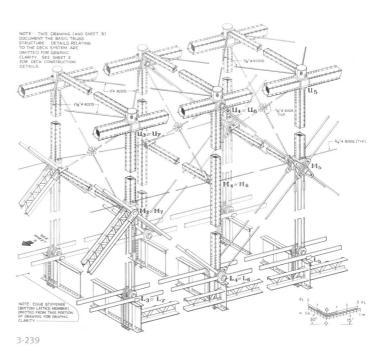

3-239

3-240. Elevation, Millgrove Road Bridge, Little Miami River, Morrow Vic., Warren County, Ohio, 1892. Toledo Bridge Company, builder. Benjamin G. Bell, delineator, 1984. P&P,HAER,OHIO,83-MOR.V,1-,sheet no. 1.

3-241. Interior view, Millgrove Road Bridge, Little Miami River, Morrow Vic., Warren County, Ohio, 1892. Toledo Bridge Company, builder. Paul Cromer, photographer, 1984. P&P,HAER,OHIO,83-MOR.V,1-8.

3-242. Post Road Bridge, spanning former Pennsylvania Railroad tracks, Havre de Grace Vic., Harford County, Maryland, 1905. American Bridge Company (Pittsburgh, PA), builder. Alvin MacDonald, photographer, 1982. P&P,HAER,MD,13-HAV,3-5.

3-243. Narrowsburg Bridge and toll house, Delaware River, Narrowsburg, Sullivan County, New York, 1899 (replaced 1953). Oswego Bridge Company, builder. Unidentified photographer and date. P&P,LC-USZ62-83460.

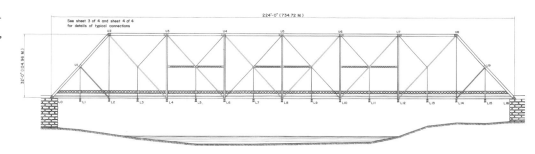

SOUTH ELEVATION
Scale: 1/8" = 1'-0"

3-240

3-241

3-243

3-242

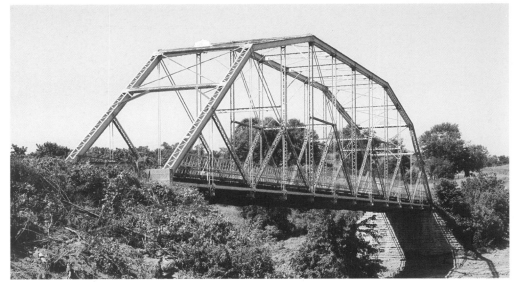
3-244

3-245

3-246

Pennsylvania Truss

The Pennsylvania, or Petit, subdivided truss was developed by the Pennsylvania Railroad in 1875. Like the Parker truss, its upper chords are arranged as a polygon.

3-244. Starnes Bridge, Eagle Creek, Holbrook Vic., Grant County, Kentucky, 1890. King Iron Bridge & Manufacturing Company (Cleveland, OH), builder. G. D. Rawlings and J. E. Daniel II, photographers, 1983. P&P,HAER,KY,41-HOLB.V,1-5.

3-245. Old St. Charles Bridge (Old Route 40 Bridge), Missouri River, St. Charles, St. Charles County, Missouri, 1902–1904 (demolished 1997). J. A. L. Waddell, engineer; Midland Bridge Company (Kansas City, MO), builder. James Rouch, photographer, 1988. P&P,HAER,MO,92-SAICH,31-10.

3-246. Traer Street Bridge (Greene Bridge), Shell Rock River, Greene, Butler County, Iowa, 1902–1903 (demolished 1981). Clinton Bridge & Iron Works (Clinton, IA), builder. Robert Ryan & Martha Bowers, photographers, 1980. P&P,HAER,IOWA,12-GREENE,1-20.

This 250-foot, steel pin-connected span played an important role in the economic life of Greene by linking the town's shops and grain elevators to the farms on the other side of the river.

3-247. Traer Street Bridge (Greene Bridge), Shell Rock River, Greene, Butler County, Iowa, 1902–1903 (demolished 1981). Clinton Bridge & Iron Works (Clinton, IA), builder. Robert Ryan & Martha Bowers, photographers, 1980. P&P,HAER,IOWA,12-GREENE,1-5.

3-248. Sidewalk, Traer Street Bridge (Greene Bridge), Shell Rock River, Greene, Butler County, Iowa, 1902–1903 (demolished 1981). Clinton Bridge & Iron Works (Clinton, IA), builder. Robert Ryan & Martha Bowers, photographers, 1980. P&P,HAER,IOWA,12-GREENE,1-19.

3-249. Pinned connections, Traer Street Bridge (Greene Bridge), Shell Rock River, Greene, Butler County, Iowa, 1902–1903 (demolished 1981). Clinton Bridge & Iron Works (Clinton, IA), builder. Robert Ryan & Martha Bowers, photographers, 1980. P&P,HAER,IOWA,12-GREENE,1-12.

3-247

3-248

3-249

3-250

3-251

3-252

3-250. Vandalia Bridge, Milk River, Glasgow Vic., Valley County, Montana, 1910. Central States Bridge Company (Indianapolis, IN), builder. Paul Anderson, photographer, 1988. P&P,HAER,MONT,53-GLA.V,1-3.

3-251. Manzanola Bridge, Arkansas River, Manzanola Vic., Crowley County, Colorado, 1911 (originally spanned the Colorado River at Clifton; moved 1950; demolished 1984). Patterson-Burghardt Bridge Company (Denver, CO), builder. Clayton Fraser, photographer, 1984. P&P,HAER,COLO,13-MANZ.V, 1-3.

3-252. New Geneva Bridge, Monongahela Railroad, Monongahela River, New Geneva Vic., Fayette County, Pennsylvania, 1912–1913. J. C. Bland, engineer; American Bridge Company (Pittsburgh, PA), builder. Jet T. Lowe, photographer, 1999. P&P,HAER,PA,26-NEGEN.V,2-2.

Upon completion, this 475-foot, steel pin-connected span was heralded for its erection using the end-launching technique, by which the truss was assembled on one bank and slid across the river supported by scaffolding mounted on a barge. End launching itself was not new (3-255). Engineers routinely used the technique for smaller bridges, but its application for such a long span weighing 1,800 tons was notable.

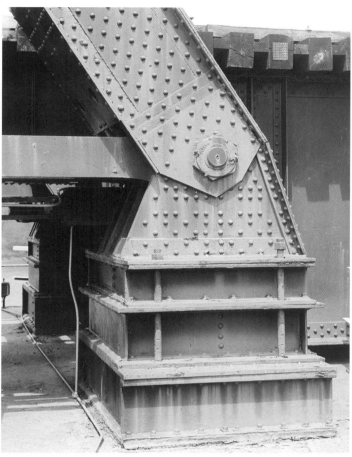

3-253

3-254

3-253. New Geneva Bridge, Monongahela Railroad, Monongahela River, New Geneva Vic., Fayette County, Pennsylvania, 1912–1913. J. C. Bland, engineer; American Bridge Company (Pittsburgh, PA), builder. Jet T. Lowe, photographer, 1999. P&P,HAER,PA,26-NEGEN.V,2-7.

3-254. End pivot bearing, New Geneva Bridge, Monongahela Railroad, Monongahela River, New Geneva Vic., Fayette County, Pennsylvania, 1912–1913. J. C. Bland, engineer; American Bridge Company (Pittsburgh, PA), builder. Jet T. Lowe, photographer, 1999. P&P,HAER,PA,26-NEGEN.V,2-10.

3-255. "Champion's Mode of Building and Transporting Bridges," patented by Thomas and Samuel Champion, 1853. Illustrated broadsheet published in Washington, D.C., 1853. Printed Ephemera Collection, Portfolio 201, Folder 13.

These engravings illustrate a system for end launching an arch-reinforced truss. The span would be assembled on tracked rollers, slid across the near abutment to a barge, and winched across the river to the far abutment.

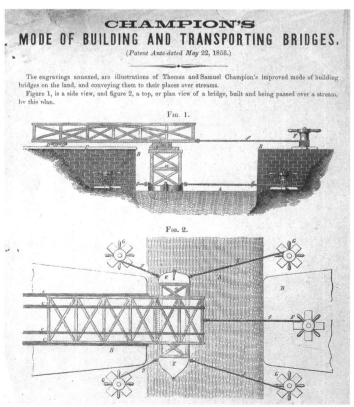

3-255

3-256

3-257

3-256. Irigary Bridge, Powder River, Sussex Vic., Johnson County, Wyoming, 1913. Canton Bridge Company (Canton, OH), builder. Clayton Fraser, photographer, 1982. P&P,HAER,WYO,10-SUS.V, 1-1.

3-257. Old Brownsville Bridge, Monongahela River, Brownsville, Fayette County, Pennsylvania, 1914. George Porter (Fayette County), Chaney, Armstrong (Washington County), engineers; Crossan Construction Company (Philadelphia, PA), substructure; Fort Pitt Bridge Works (Canonsburg, PA), superstructure. Joseph Elliott, photographer, 1997. P&P,HAER,PA,26-BROVI,5-2.

3-258. North Carolina Route 126 Bridge, Lake James Canal, Marion Vic., Burke County, North Carolina, 1919. Virginia Bridge & Iron Company (Roanoke, VA), builder. Rick Lanier, photographer, 1984. P&P,HAER,NC,12-MAR.V,1-1.

This 300-foot steel bridge has more braces and shorter, heavier members than a typical Pennsylvania truss of its size and load-bearing capacity due to dimensional restrictions of materials trucked over the mountainous, unimproved roads, which were the only access to the site.

3-259. Detail of panels, North Carolina Route 126 Bridge, Lake James Canal, Marion Vic., Burke County, North Carolina, 1919. Virginia Bridge & Iron Company (Roanoke, VA), builder. Rick Lanier, photographer, 1984. P&P,HAER,NC,12-MAR.V,1-4.

3-260. McGirt's Bridge, Cape Fear River, Elizabethtown Vic., Bladen County, South Carolina, 1923 (demolished). Atlantic Bridge Company (Charlotte, NC), builder; Virginia Bridge & Iron Company (Roanoke, VA), truss fabricator. Randell Page, photographer, 1983. P&P,HAER,NC,9-ELITO.V,1-4.

3-261

3-262

3-263

3-261. Four Mile Bridge, Big Horn River, Thermopolis Vic., Hot Springs County, Wyoming, 1927–1928. Charles M. Smith (Thermopolis, WY), builder. Clayton Fraser, photographer, 1982. P&P,HAER,WYO,9-THERM.V,1-1.

3-262. Big Four Bridge, Cleveland, Cincinnati, Chicago & St. Louis Railroad, Ohio River, Louisville, Kentucky, 1929. L. H. Schaeperklaus, W. S. Burnett, engineers; Louisville & Jefferson Bridge Company, builder. Jack E. Boucher, photographer, 1975. P&P,HAER,KY,56-LOUVI,71-3.

3-263. Bridgeport Bridge, Wisconsin River, Bridgeport, Crawford County, Wisconsin, 1931 (demolished 1990). C. H. Kirsch, State Highway Commission, designer; Schaffer Construction Company, builder; Steve Brothers, fabricator. Mark R. Ray, photographer, 1987. P&P,HAER,WIS,12-BRIDG,1-5.

WHIPPLE-MURPHY TRUSS
(DOUBLE-INTERSECTION PRATT TRUSS)

Also known as the Linville truss, this variation of the Pratt truss features diagonal members that extend across two panels. The basic configuration, which increases the rigidity of the truss, was designed by Squire Whipple in 1847 and soon after improved upon by John W. Murphy, chief engineer of the Lehigh Valley Railroad, and Jacob H. Linville, chief engineer of the Pennsylvania Railroad who later became president of the Keystone Bridge Company.

3-264. Riverside Avenue Bridge, spanning Amtrak Northeast Corridor right-of-way, Greenwich, Fairfield County, Connecticut, 1871 (originally spanned the Housatonic River in Stratford; moved ca. 1894). F. C. Lowthorp, engineer, Keystone Bridge Company (Pittsburgh, PA), builder. Jet T. Lowe, photographer, 1984. P&P,HAER,CONN,1-GREWI,1-7.

This 164-foot span composed of cast- and wrought-iron pin-connected members was built originally as part of a six-span bridge crossing the Housatonic River in Stratford.

3-265. Mead Avenue Bridge, French Creek, Meadeville, Crawford County, Pennsylvania, 1871, 1912. Penn Bridge Company (New Brighton, PA, 1871), Rodgers Bros. Company (Albion, PA, 1912), builders. Jack E. Boucher, photographer, 1972. P&P,HAER,PA, 20-MEDVI,6-6.

Increased traffic loads led to the encasement of the original two-span, wrought-iron Whipple-Murphy trusses with steel Baltimore trusses in 1912.

3-264

3-265

3-266

3-266. Girard Avenue Bridge, Schuylkill River, Philadelphia, Pennsylvania, 1873 (demolished 1971). Phoenix Bridge Company (Phoenixville, PA), builder. Colored etching, 1877. P&P Vertical File, 4384-G.

3-267. Detail, Girard Avenue Bridge, Schuylkill River, Philadelphia, Pennsylvania, 1873 (demolished 1971). Phoenix Bridge Company (Phoenixville, PA), builder. Unidentified photographer, 1969 or 1971. P&P,HABS,PA,51-PHILA,461-10.

3-267

3-268. Howard Bridge, Kokosing River, Howard, Knox County, Ohio, 1874. David H. and Charles C. Morrison, Columbia Bridge Company (Dayton, OH), designers and builder. Joseph Elliott, photographer, 1992. P&P,HAER,OHIO,42-HOW,1-1.

The Morrisons were prominent bridge builders in Ohio. This 96-foot span is an example of their version of the double-intersection Pratt truss, which they marketed under the names "Morrison's Iron Patented Wrought Iron Truss Bridge" and "Quadrilateral Truss Bridge."

3-269. Connection details at deck, Howard Bridge, Kokosing River, Howard, Knox County, Ohio, 1874. David H. and Charles C. Morrison, Columbia Bridge Company (Dayton, OH), designers and builder. Joseph Elliott, photographer, 1992. P&P,HAER,OHIO,42-HOW,1-11.

3-270. Detail of bearing, Howard Bridge, Kokosing River, Howard, Knox County, Ohio, 1874. David H. and Charles C. Morrison, Columbia Bridge Company (Dayton, OH), designers and builder. Joseph Elliott, photographer, 1992. P&P,HAER,OHIO,42-HOW,1-12.

3-268

3-269

3-270

3-271

3-272

3-273

3-271. Eveland Bridge, Des Moines River, Oskaloosa Vic., Mahaska County, Iowa, 1876–1877. C. W. Tracy, engineer. McKay & Nelson (Fort Wayne, IN), builders. Joseph Elliott, photographer, 1995. P&P,HAER,IOWA,62-OSK.V,2-2.

3-272. Chicago & Alton Railway Bridge, Missouri River, Glasgow, Howard County, Missouri, 1878–1879 (demolished 1922). William Sooy Smith, engineer; Abram T. Hay, steel specifications; Carnegie Steel Company (Pittsburgh, PA), steel manufacturer. Unidentified photographer, ca. 1880. P&P,HABS,MO,45-GLASG,4A-1.

The first all-steel bridge erected in the United States consisted of five Whipple-Murphy spans each measuring 314 feet in length. William Sooy Smith (1830–1916) achieved prominence as a structural engineer and as a Union cavalry commander during the Civil War. His engineering practice after the war included the design of innovative foundations and steel frames for tall buildings in Chicago as well as bridges.

3-273. Hardin City Bridge, Steamboat Rock Vic., Hardin County, Iowa, 1879 (originally spanned Iowa River; moved to present site alongside county road 1989). Western Bridge Works (Fort Wayne, IN), builder; Passaic Rolling Mills (Passaic, NJ), fabricator. Joseph Elliott, photographer, 1995. P&P,HAER,IOWA,42-STERO.V,1-2.

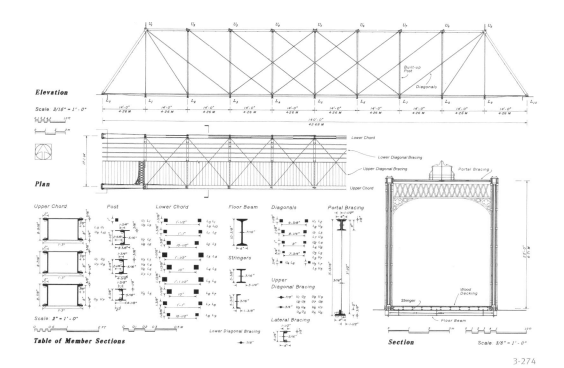

3-274

3-274. Plan, elevation, section, and table of member sections, Hardin City Bridge, Steamboat Rock Vic., Hardin County, Iowa, 1879 (originally spanned Iowa River; moved to present site alongside county road 1989). Western Bridge Works (Fort Wayne, IN), builder; Passaic Rolling Mills (Passaic, NJ), fabricator. Caroline Schweyer and Adriaan Vlaardingerbroek, delineators, 1995. P&P,HAER,IOWA,42-STERO.V,1-,sheet no. 2.

3-275. Detail of bearing, Hardin City Bridge, Steamboat Rock Vic., Hardin County, IA, 1879 (originally spanned Iowa River; moved to present site alongside county road 1989). Western Bridge Works (Fort Wayne, IN), builder; Passaic Rolling Mills (Passaic, NJ), fabricator. Joseph Elliott, photographer, 1995. P&P,HAER,IOWA,42-STERO.V,1-10.

3-276. Connection details at deck, Hardin City Bridge, Steamboat Rock Vic., Hardin County, Iowa, 1879 (originally spanned Iowa River; moved to present site alongside county road 1989). Western Bridge Works (Fort Wayne, IN), builder; Passaic Rolling Mills (Passaic, NJ), fabricator. Joseph Elliott, photographer, 1995. P&P,HAER,IOWA,42-STERO.V,1-9.

3-275

3-276

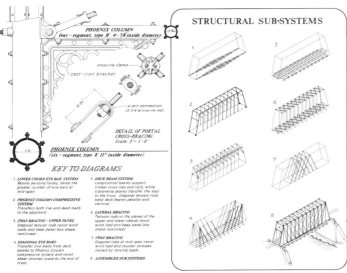

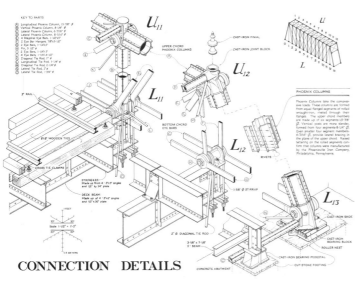

CONNECTION DETAILS

3-277. Hayden Bridge, Central Pacific Railroad, McKenzie River, Springfield Vic., Lane County, Oregon, 1882 (originally spanned Bear River, near Corrine, UT; moved to present location 1900). Clarke, Reeves & Company, Phoenix Bridge Company (Phoenixville, PA), designer and fabricator. Jet T. Lowe, photographer, 1990. P&P,HAER,ORE,20-SPRIF,2-2.

3-278. Diagrams of structural subsystems, Hayden Bridge, Central Pacific Railroad, McKenzie River, Springfield Vic., Lane County, Oregon, 1882 (originally spanned Bear River, near Corrine, UT; moved to present location 1900). Clarke, Reeves & Company, Phoenix Bridge Company (Phoenixville, PA), designer and fabricator. Todd A. Croteau, Richard L. Koochagian, Gretchen Van Dusen, Rafael Villalobos, delineators, 1990. P&P,HAER,ORE,20-SPRIF,2-,sheet no. 4.

3-279. Connection details, Hayden Bridge, Central Pacific Railroad, McKenzie River, Springfield Vic., Lane County, Oregon, 1882 (originally spanned Bear River, near Corrine, UT; moved to present location 1900). Clarke, Reeves & Company, Phoenix Bridge Company (Phoenixville, PA), designer and fabricator. Todd A. Croteau, Richard L. Koochagian, Gretchen Van Dusen, Rafael Villalobos, delineators, 1990. P&P,HAER,ORE,20-SPRIF,2-,sheet no. 3.

3-280. Interior view, Hayden Bridge, Central Pacific Railroad, McKenzie River, Springfield Vic., Lane County, Oregon, 1882 (originally spanned Bear River, near Corrine, UT; moved to present location 1900). Clarke, Reeves & Company, Phoenix Bridge Company (Phoenixville, PA), designer and fabricator. Jet T. Lowe, photographer, 1990. P&P,HAER,ORE,20-SPRIF,2-4.

3-281

3-282

3-283

3-281. Phoenix column and pinned bearing on roller nest, Hayden Bridge, Central Pacific Railroad, McKenzie River, Springfield Vic., Lane County, Oregon, 1882 (originally spanned Bear River, near Corrine, UT; moved to present location 1900). Clarke, Reeves & Company, Phoenix Bridge Company (Phoenixville, PA), designer and fabricator. Jet T. Lowe, photographer, 1990. P&P,HAER,ORE,20-SPRIF,2-12.

The bearings at one end of the span rest on a row of small-diameter rollers that allow the truss to expand and contract in response to changing loads and thermal conditions.

3-282. Lower chord connections at deck beam, Hayden Bridge, Central Pacific Railroad, McKenzie River, Springfield Vic., Lane County, Oregon, 1882 (originally spanned Bear River, near Corrine, UT; moved to present location 1900). Clarke, Reeves & Company, Phoenix Bridge Company (Phoenixville, PA), designer and fabricator. Jet T. Lowe, photographer, 1990. P&P,HAER,ORE,20-SPRIF,2-8.

3-283. Lower chord connections at post, Hayden Bridge, Central Pacific Railroad, McKenzie River, Springfield Vic., Lane County, Oregon, 1882 (originally spanned Bear River, near Corrine, UT; moved to present location 1900). Clarke, Reeves & Company, Phoenix Bridge Company (Phoenixville, PA), designer and fabricator. Jet T. Lowe, photographer, 1990. P&P,HAER,ORE,20-SPRIF,2-10.

George S. Morison's Missouri River Bridges

In the decades following the Civil War, engineers such George Shattuck Morison (1842–1903) played important supporting roles in the fierce compeition among railroad companies establishing routes west of the Mississippi River. Trained in law and admitted to the bar, Morison found his calling in engineering under the tutelage of Octave Chanute (see 4-003). He achieved prominence as a bridge designer and became a leader in the emergence of engineering as a learned profession. For the Burlington Railroad and its allies, Morison created a standard design using Whipple-Murphy trusses for seven bridges crossing the Missouri River. These spans mark the transition in American engineering from composite construction using cast and wrought iron—Plattsmouth (1880), Bismarck (1883)—to steel—Sioux City (1888), Nebraska City (1888), Rulo (1889).

3-284. Plattsmouth Bridge, Chicago, Burlington & Quincy Railroad, Missouri River, Plattsmouth Vic., Cass County, Nebraska, 1880 (demolished). George S. Morison, chief engineer; Keystone Bridge Company (Pittsburgh, PA), Kellogg & Maurice (Athens, PA), superstructure; Reynolds, Saulpaugh & Company (Rock Island, IL), masonry. Unidentified photographer, ca. 1880, from George S. Morison, The Plattsmouth Bridge, 1882. P&P,HAER,NEB,13-PLATT.V,1-1.

3-285. Section showing construction of pier and pneumatic caisson, Plattsmouth Bridge, Chicago, Burlington & Quincy Railroad, Missouri River, Plattsmouth Vic., Cass County, Nebraska, 1880 (demolished). George S. Morison, chief engineer; Keystone Bridge Company (Pittsburgh, PA), Kellogg & Maurice (Athens, PA), superstructure; Reynolds, Saulpaugh & Company (Rock Island, IL), masonry. Unidentified delineator, ca. 1880, from George S. Morison, The Plattsmouth Bridge, 1882. P&P,HAER,NEB,13-PLATT.V,1-6.

Pneumatic caissons for underwater excavation were first used in the United States in 1870 by James Eads and John Roebling in the construction of their respective bridges in St. Louis and New York City (see 2-050, 5-035).

3-286. Bismarck Bridge, Northern Pacific Railroad, Missouri River, Bismarck, Burleigh County, North Dakota, 1881–1883 (demolished). George S. Morison, chief engineer; Detroit Bridge & Iron Works (Detroit, MI), superstructure; T. Saulpaugh & Company (Rock Island, IL), substructure. Unidentified photographer, from George S. Morison, The Bismarck Bridge, 1884. P&P,HAER,ND,8-BISMA,3-1.

3-287. Nebraska City Bridge, Chicago, Burlington & Quincy Railroad, Missouri River, Nebraska City Vic., Otoe County, Nebraska, 1887–1888 (demolished). George S. Morison, chief engineer; Union Bridge Company (Athens, PA), superstructure fabrication; Baird Brothers (Pittsburgh, PA), superstructure erection; T. Saulpaugh & Company (Rock Island, IL), substructure. Unidentified photographer, from George S. Morison, The Nebraska City Bridge, 1892. P&P,HAER,NEB,66-NEBCI,5-47.

3-288. Map of site, Nebraska City Bridge, Chicago, Burlington & Quincy Railroad, Missouri River, Nebraska City Vic., Otoe County, Nebraska, 1887–1888 (demolished). George S. Morison, chief engineer; Union Bridge Co. (Athens, PA), superstructure fabrication; Baird Brothers (Pittsburgh, PA), superstructure erection; T. Saulpaugh & Co., substructure. Robert A. Welcke, photolithographer, from George S. Morison, The Nebraska City Bridge, 1892. P&P,HAER, NEB,66-NEBCI,5-48.

3-286

3-287

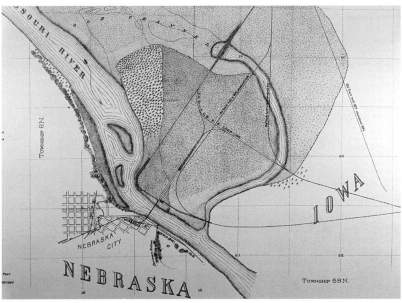

3-288

3-289

3-290

3-289. Deck truss approach, Nebraska City Bridge, Chicago, Burlington & Quincy Railroad, Missouri River, Nebraska City Vic., Otoe County, Nebraska, 1887–1888 (demolished). George S. Morison, chief engineer; Union Bridge Company (Athens, PA), superstructure fabrication; Baird Brothers (Pittsburgh, PA), superstructure erection; T. Saulpaugh & Company (Rock Island, IL), substructure. Clayton Fraser, photographer, 1984. P&P,HAER,NEB, 66-NEBCI,5-16.

3-290. Elevations of masonry pier supported on wooden piles, Nebraska City Bridge, Chicago, Burlington & Quincy Railroad, Missouri River, Nebraska City Vic., Otoe County, Nebraska, 1887–1888 (demolished). George S. Morison, chief engineer; Union Bridge Company (Athens, PA), superstructure fabrication; Baird Brothers (Pittsburgh, PA), superstructure erection; T. Saulpaugh & Company (Rock Island, IL), substructure. From George S. Morison, The Nebraska City Bridge, 1892. P&P,HAER, NEB,66-NEBCI,5-52.

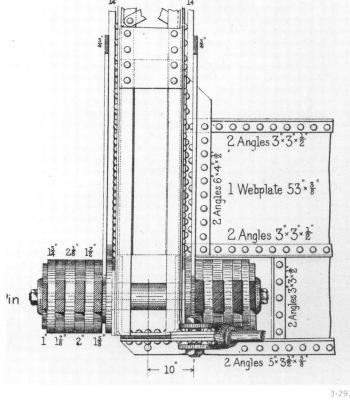

3-291

3-291. Interior, Nebraska City Bridge, Chicago, Burlington & Quincy Railroad, Missouri River, Nebraska City Vic., Otoe County, Nebraska, 1887–1888 (demolished). George S. Morison, chief engineer; Union Bridge Company (Athens, PA), superstructure fabrication; Baird Brothers (Pittsburgh, PA), superstructure erection; T. Saulpaugh & Company (Rock Island, IL), substructure. Clayton Fraser, photographer, 1984. P&P,HAER,NEB,66-NEBCI,5-20.

3-292. Connection details, bottom chord and deck beam, Nebraska City Bridge, Chicago, Burlington & Quincy Railroad, Missouri River, Nebraska City Vic., Otoe County, Nebraska, 1887–1888 (demolished). George S. Morison, chief engineer; Union Bridge Company (Athens, PA), superstructure fabrication; Baird Brothers (Pittsburgh, PA), superstructure erection; T. Saulpaugh & Company (Rock Island, IL), substructure. Robert A. Welcke, photo-lithographer, from George S. Morison, The Nebraska City Bridge, 1892. P&P,HAER, NEB,66-NEBCI,5-58.

3-293. Connection details, bottom chord, vertical, and deck beam, Nebraska City Bridge, Chicago, Burlington & Quincy Railroad, Missouri River, Nebraska City Vic., Otoe County, Nebraska, 1887–1888 (demolished). George S. Morison, chief engineer; Union Bridge Company (Athens, PA), superstructure fabrication; Baird Brothers (Pittsburgh, PA), superstructure erection; T. Saulpaugh & Company (Rock Island, IL), substructure. Clayton Fraser, photographer, 1984. P&P,HAER,NEB,66-NEBCI,5-31.

3-293

3-294

3-295

3-294. Sioux City Railroad Bridge, Chicago & North Western Railroad, Missouri River, Sioux City, Woodbury County, Iowa, 1888 (demolished). George S. Morison, chief engineer; Union Bridge Company (Athens, PA), superstructure fabrication; Baird Brothers (Pittsburgh, PA), superstructure erection; T. Saulpaugh (Rock Island, IL), masonry. Unidentified photographer, from George S. Morison, The Sioux City Bridge, 1890. P&P,HAER,IA-96-1.

3-295. Rulo Bridge, Chicago, Burlington & Quincy Railroad, Missouri River, Rulo, Richardson County, Nebraska, 1889 (demolished). George S. Morison, chief engineer; Edge Moor Iron Company (Edgemoor, DE), superstructure fabrication; Baird Brothers (Pittsburgh, PA), superstructure erection; Drake & Stratton, masonry. Unidentified photographer, from George S. Morison, The Rulo Bridge, 1890. P&P,HAER,NEB,74-RULO,1-10.

TRIPLE INTERSECTION PRATT TRUSS

Rarely built in the United States, triple intersection Pratt trusses extend the diagonals across three panels.

3-296

3-296. Laughery Creek Bridge, Aurora Vic., Dearborn County, Indiana, 1878. Wrought Iron Bridge Company (Canton, OH), builder. Jack E. Boucher, photographer, 1974. P&P,HAER,IND,15-AUR.V,1-2.

This 300-foot bridge is the only triple intersection Pratt truss known to be extant in the United States.

3-297. Truss detail, Laughery Creek Bridge, Aurora Vic., Dearborn County, Indiana, 1878. Wrought Iron Bridge Company (Canton, OH), builder. Jack E. Boucher, photographer, 1974. P&P,HAER,IND,15-AUR.V,1-9.

3-297

Wendel Bollman (1814–1884) began his career as a carpenter and learned bridge design
while working for the Baltimore & Ohio Railroad under Bejamin Latrobe II (see 2-031–
2-033). Around 1850, he developed a truss that superimposed a suspension system consisting of long, wrought-iron bars extending from the end posts to each panel point upon
a post-and-diagonal panel system recalling a Pratt truss, but lacking a true bottom chord.
The B&O used it for about twenty years. Only one example is known to remain today.

3-298

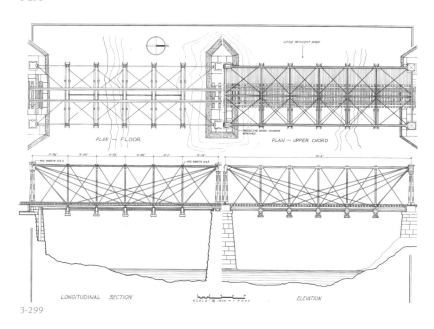

3-299

3-298. Bollman Truss Bridge, Baltimore & Ohio Railroad, Little
Patuxent River, Savage, Howard County, Maryland, 1869 (moved
to present site from main line 1887; restored 1983–1984). Wendel Bollman, engineer. Jet T. Lowe, photographer, 1996.
P&P,HAER,MD,14-SAV,1-13.

3-299. Plans, elevation, and section, Bollman Truss Bridge, Balti-
more & Ohio Railroad, Little Patuxent River, Savage, Howard
County, Maryland, 1869 (moved to present site from main line
1887; restored 1983–1984). Wendel Bollman, engineer. Gregory
Brezinski and Jeffrey Jenkins, delineators, 1970.
P&P,HAER,MD,14-SAV,1-,sheet no. 2.

3-300. Interior, Bollman Truss Bridge, Baltimore & Ohio
Railroad, Little Patuxent River, Savage, Howard County,
Maryland, 1869 (moved to present site from main line
1887; restored 1983–1984). Wendel Bollman, engi-
neer. William E. Barrett, photographer, 1970.
P&P,HAER,MD,14-SAV,1-10.

3-301. Isometric view of connection at end panel, Bollman Truss Bridge, Baltimore & Ohio Railroad, Little
Patuxent River, Savage, Howard County, Maryland,
1869 (moved to present site from main line 1887;
restored 1983–1984). Wendel Bollman, engineer.
Michael Masny, delineator, 1970. P&P,HAER,MD,14-
SAV,1-, sheet no. 7.

3-302. End post, Bollman Truss Bridge, Baltimore &
Ohio Railroad, Little Patuxent River, Savage, Howard
County, Maryland, 1869 (moved to present site from
main line 1887; restored 1983–1984). Wendel Boll-
man, engineer. Jet T. Lowe, photographer, 1996.
P&P,HAER,MD,14-SAV,1-15.

3-300

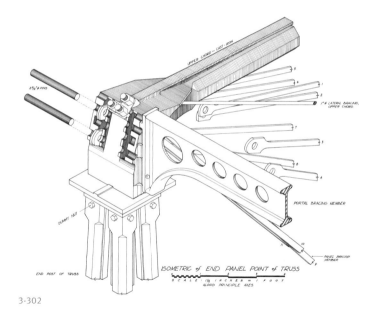

3-302

3-301

3-303

3-303. Connections at deck beam, Bollman Truss Bridge, Balti-more & Ohio Railroad, Little Patuxent River, Savage, Howard County, Maryland, 1869 (moved to present site from main line 1887; restored 1983–1984). Wendel Bollman, engineer. William E. Barrett, photographer, 1970. P&P,HAER,MD,14-SAV,1-5.

The perforated members between the deck beams are spacers working in compression rather than the typical bottom chord in tension of a Pratt truss.

3-304. Connections at lower chord and deck beam, Bollman Truss Bridge, Baltimore & Ohio Railroad, Little Patuxent River, Savage, Howard County, Maryland, 1869 (moved to present site from main line 1887; restored 1983–1984). Wendel Bollman, engineer. Charles Parrott III, delineator, 1970. P&P,HAER,MD,14-SAV,1-,sheet no. 4.

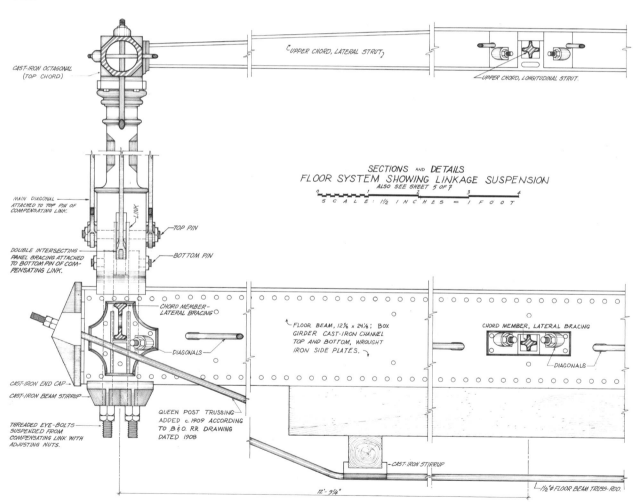

3-304

FINK TRUSS

After studying architecture and engineering at the polytechnical institute in Darmstadt, Germany, Albert Fink (1827–1897) immigrated to the United States in 1849 and became chief assistant to Benjamin Latrobe II (see 2-031–2-033) in the engineering department at the Baltimore & Ohio Railroad, where Wendel Bollman (3-298–3-304) also worked. In the 1850s, Fink developed a truss that was similar to Bollman's in its combination of panel and suspension features, but it had a simpler and more economical layout with symmetrical diagonals. In its deck truss configuration, it is distinctive for the absence of a bottom chord. Fink trusses were commonly used for bridges through the 1870s, but only one example is known to be extant today.

3-305

3-305. Fairmont Bridge, Baltimore & Ohio Railroad, Monongahela River, Fairmont, Marion County, West Virginia, 1852 (demolished). Albert Fink, engineer. Engraving from The New World, or the United States and Canada, ca. 1858. P&P,HAER,WVA,25-FAIR.V,1-1.

Fink's first major bridge consisted of three 205-foot spans; it was the longest iron bridge in the United States at the time of completion.

3-306. Missouri River railroad bridge, Wabash Railroad Company, St. Charles, St. Charles County, Missouri, 1868–1871 (demolished 1936). C. Shaler Smith, engineer. Undated engraving. P&P,LC-USZ62-29714.

3-306

3-307

3-307. Raritan River Fink through-truss bridge (Hamden Bridge), Hamden, Hunterdon County, New Jersey, 1857–1858 (demolished 1978, salvaged members moved to Hunterdon County Government Complex, Flemington Vic.). Trenton Locomotive & Machine Works (Trenton, NJ), fabricator. Jack E. Boucher, photographer, 1971. P&P,HAER,NJ,10-CLIN.V,1-7.

3-308. Connections at portal, Raritan River Fink through-truss bridge (Hamden Bridge), Hamden, Hunterdon County, New Jersey, 1857–1858 (demolished 1978, salvaged members moved to Hunterdon County Government Complex, Flemington Vic.). Trenton Locomotive & Machine Works (Trenton, NJ), fabricator. Jack E. Boucher, photographer, 1971. P&P,HAER,NJ,10-CLIN.V,1-12.

3-308

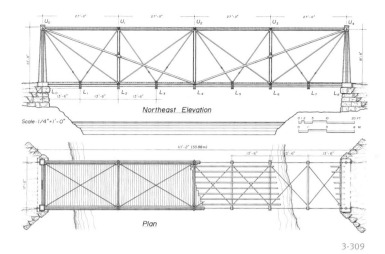

3-309

3-309. Plan and elevation, Zoarville Station Bridge, Conotton Creek, Zoarville Vic., Tuscarawas County, Ohio, 1868 (originally part of a three-span bridge over the Tuscarawas River in Dover, OH; moved to present site 1905; dismantled for restoration 2000). Smith, Latrobe & Company (Baltimore, MD), builders. Chris Payne, delineator, 1992. P&P,HAER,OHIO,79-ZOARV.V,1-,sheet no. 2.

3-310. Zoarville Station Bridge, Conotton Creek, Zoarville Vic., Tuscarawas County, Ohio, 1868 (originally part of a three-span bridge over the Tuscarawas River in Dover, OH; moved to present site 1905; dismantled for restoration 2000). Smith, Latrobe & Company (Baltimore, MD), builders. Joseph Elliott, photographer, 1992. P&P,HAER,OHIO,79-ZOARV.V,1-1.

3-311. Detail of panel, Zoarville Station Bridge, Conotton Creek, Zoarville Vic., Tuscarawas County, Ohio, 1868 (originally part of a three-span bridge over the Tuscarawas River in Dover, OH; moved to present site 1905; dismantled for restoration 2000). Smith, Latrobe & Company (Baltimore, MD), builders. Joseph Elliott, photographer, 1992. P&P,HAER,OHIO,79-ZOARV.V,1-10.

3-310

3-311

3-312

3-313

CONNECTION DETAILS

Scale: 1-1/2" = 1'-0"

U_2

Upper Chord

Cast-Iron Joint Box

2" Ø Pin

U_3

Lateral Strut

L_4

Foot Box with 3" Ø Pin

Eye Bar Hanger

Link Plate with 3" Ø Pins

7" Stringer

2" Ø Pin

3/4" Foot Plate

Adjustment Nut

L_5

Loop Bar 1-1/2" x 1"

Upper Chord

L_6

6" Ø Pin

U_4

1/2" Washer Plate

Floor Beam 15" x 5-1/2" I-beam

Cast-Iron Foot Box with 2" Ø Pin

FEET 1.0

Scale : 1 1/2" = 1'-0"

METERS 0.5

(A) End Tower 8" Ø Phoenix Column with 1/4" splice plate & 2-1/2', 2-1/2"x 1/4" angles.

(B) Longitudinal 8" Ø Phoenix Col.

(C) Vertical 6" Ø Phoenix Column

(D) Two Diagonal Eye-bars 4" x 3/4"

(E) 2 Diagonal Eye-bars 2"x 1/2"

(F) Two Diagonal Eye-bars 1-3/4" x 1/2"

(G) Two Tees 5" x 2-1/2" x 3/8"

(H) Lateral Diagonal Tie Rod 1-1/4" Ø

(I) Longitudinal Tie Rod 1-1/4" Ø

3-314

3-312. Detail of end post and portal, Zoarville Station Bridge, Conotton Creek, Zoarville Vic., Tuscarawas County, Ohio, 1868 (originally part of a three-span bridge over the Tuscarawas River in Dover, OH; moved to present site 1905; dismantled for restoration 2000). Smith, Latrobe & Company (Baltimore, MD), builders. Joseph Elliott, photographer, 1992. P&P,HAER,OHIO,79-ZOARV.V,1-6.

3-313. Detail of connections at deck beam, Zoarville Station Bridge, Conotton Creek, Zoarville Vic., Tuscarawas County, Ohio, 1868 (originally part of a three-span bridge over the Tuscarawas River in Dover, OH; moved to present site 1905; dismantled for restoration 2000). Smith, Latrobe & Company (Baltimore, MD), builders. Joseph Elliott, photographer, 1992. P&P,HAER,OHIO,79-ZOARV.V,1-16.

3-314. Connection details, Zoarville Station Bridge, Conotton Creek, Zoarville Vic., Tuscarawas County, Ohio, 1868 (originally part of a three-span bridge over the Tuscarawas River in Dover, OH; moved to present site 1905; dismantled for restoration 2000). Smith, Latrobe & Company (Baltimore, MD), builders. Chris Payne, delineator, 1992. P&P,HAER,OHIO,79-ZOARV.V,1-,sheet no. 3.

POST TRUSS

Like Wendel Bollman, Simeon S. Post (1805–1872) was trained as a carpenter and made his career as a self-taught railroad engineer. In the mid-1860s, he invented a truss in which both the compression and tension members are inclined, and he also patented a system for connections that especially appealed to the builders of railroad bridges for their rigidity. Only three bridges of his design are known to be extant.

3-315. Atherton Bridge, Nashua River, Lancaster Vic., Worcester County, Massachusetts, 1870. J. H. Cofrode & Company (Philadelphia, PA), builder. Jet T. Lowe, photographer, 1979. P&P,HAER,MASS,14-LANC.V,1-2.

This 72-foot pony truss highway bridge employs the web configuration of the Post truss without Post's patented joint connections.

3-316. Ponakin Road Bridge, Nashua River, Lancaster Vic., Worcester County, Massachusetts, 1871. Watson Manufacturing Company (Paterson, NJ), builder. Jet T. Lowe, photographer, 1979. P&P,HAER,MASS,14-LANC.V,2-1.

This 100-foot through truss employs both the Post web configuration and his patented connections.

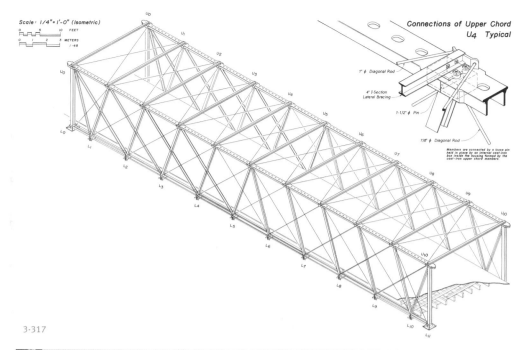

Scale: 1/4"=1'-0" (Isometric)

Connections of Upper Chord
U4 Typical

1" φ Diagonal Rod

4" I-Section
Lateral Bracing

1-1/2" φ Pin

7/8" φ Diagonal Rod

Members are connected by a loose pin
held in place by an internal cast-iron
box inside the housing formed by the
cast-iron upper chord members.

3-317

3-318

3-319

3-317. Isometric view, Ponakin Road Bridge, Nashua River, Lancaster Vic., Worcester County, Massachusetts, 1871. Watson Manufacturing Company (Paterson, NJ), builder. Pat Reese, delineator, 1990. P&P,HAER,MASS,14-LANC.V,2-,sheet no. 3.

3-318. Interior, Ponakin Road Bridge, Nashua River, Lancaster Vic., Worcester County, Massachusetts, 1871. Watson Manufacturing Company (Paterson, NJ), builder. Jet T. Lowe, photographer, 1979. P&P,HAER,MASS,14-LANC.V,2-7.

3-319. Detail of connections at upper chord, Ponakin Road Bridge, Nashua River, Lancaster Vic., Worcester County, Massachusetts, 1871. Watson Manufacturing Company (Paterson, NJ), builder. Jet T. Lowe, photographer, 1979. P&P,HAER,MASS,14-LANC.V,2-14.

Post's patented joint connectors featured a cast-iron box and sleeves into which the members were inserted and pinned.

THACHER TRUSS

Edwin Thacher (1840–1920), chief engineer of the Keystone Bridge Company, published this design for a combined panel and suspension truss in 1884. Variations were adopted by competitors, such as the Wrought Iron Bridge Company, but not built in great numbers.

3-320. Parshallburg Bridge, Shiawassee River, Chesaning Vic., Saginaw County, Michigan, 1889 (moved 1999). Wrought Iron Bridge Company (Canton, OH), builder. Dietrich Floeter, photographer, 1995. P&P,HAER,MICH,73-CHESA.V,1-1.

3-321. Thacher truss bridge, Linville Creek, Broadway, Rockingham County, Virginia, 1898. Wrought Iron Bridge Company (Canton, OH), builder. Jet T. Lowe, photographer, 1978. P&P,HAER,VA,83-BROAD,2-3.

3-322. Interior, Thacher truss bridge, Linville Creek, Broadway, Rockingham County, Virginia, 1898. Wrought Iron Bridge Company (Canton, OH), builder. Jet T. Lowe, photographer, 1978. P&P,HAER,VA,83-BROAD,2-6.

3-320

3-321

3-322

K-TRUSS

The K-truss came into use for steel-truss spans longer than 300 feet during the second decade of the twentieth century. Its webs have a distinctive K-shaped configuration formed by the intersection of a pair of diagonal braces at the midpoint of one post.

3-323. Krotz Springs Bridge, Atchafalaya River, Krotz Springs, St. Landry Parish, Louisiana, 1934. Louisiana Highway Commission, designer; Foundation Company (New York, NY), contractor; Nashville Bridge Company, fabricator. Larry Norman, photographer, 1985. P&P,HAER,LA,49-KROSP,1-4.

The Krotz Springs Bridge employs three 500-foot K-truss spans, a type favored by the Louisiana Highway Commission during the state's ambitious road improvement campaign launched by Governor Huey Long in 1930.

3-323

Patented in England in 1848 by James Warren and Theobald Willoughby Monsani, the Warren truss consists of diagonal members, alternately placed in tension and compression, forming a W pattern. Common variants on the basic design include the addition of vertical members and intersecting diagonal braces. The Warren truss lends itself to standardization and is extremely rigid when the connections are riveted or welded. As the use of fixed connections became more widely practiced in the 1920s and 1930s, it surpassed Pratt-type designs as the truss of choice for steel highway and railroad bridges.

Warren Through Truss

3-324. Wills Creek Bridge, Baltimore & Ohio Railroad, spanning CSX tracks at T381, Meyersdale Vic., Somerset County, Pennsylvania, 1871 (orginally a rail bridge spanning Wills Creek on the B&O Pittsburgh Division, moved ca. 1910). Wendel Bollman, engineer; Patapsco Bridge & Iron Works (Baltimore, MD), builder. Jet T. Lowe, photographer, 1992. P&P,HAER,PA,56-MEYER.V,2-4.

3-325. Interior view, Wills Creek Bridge, Baltimore & Ohio Railroad, spanning CSX tracks at T381, Meyersdale Vic., Somerset County, Pennsylvania, 1871 (orginally a rail bridge spanning Wills Creek on the B&O Pittsburgh Division, moved ca. 1910). Wendel Bollman, designer; Patapsco Bridge & Iron Works (Baltimore, MD), builder. Jet T. Lowe, photographer, 1992. P&P,HAER,PA,56-MEYER.V,2-5.

3-326. Wills Creek Bridge, Baltimore & Ohio Railroad, spanning CSX tracks at T381, Meyersdale Vic., Somerset County, Pennsylvania, 1871 (orginally a rail bridge spanning Wills Creek on the B&O Pittsburgh Division, moved ca. 1910). Wendel Bollman, designer; Patapsco Bridge & Iron Works (Baltimore, MD), builder. Jet T. Lowe, photographer, 1992. P&P,HAER,PA,56-MEYER.V,2-2.

3-324

3-326

3-325

3-327

3-328

3-329

3-327. Kentucky Route 2014 bridge, Cumberland River, Pineville Vic., Bell County, Kentucky, 1873 (demolished). Louisville Bridge & Iron Company (Louisville, KY), builder. J. C. Henderson, J. L. Mettille, R. M. Morris, photographers, 1987. P&P,HAER,KY,7-PINVI,2-4.

3-328. View of truss panels, Kentucky Route 2014 bridge, Cumberland River, Pineville Vic., Bell County, Kentucky, 1873 (demolished). Louisville Bridge & Iron Company (Louisville, KY), builder. J. C. Henderson, J. L. Mettille, R. M. Morris, photographers, 1987. P&P,HAER,KY,7-PINVI,2-7.

3-329. Phoenix column detail, Kentucky Route 2014 bridge, Cumberland River, Pineville Vic., Bell County, Kentucky, 1873 (demolished). Louisville Bridge & Iron Company (Louisville, KY), builder. J. C. Henderson, J. L. Mettille, R. M. Morris, photographers, 1987. P&P,HAER,KY,7-PINVI,2-10.

3-330. Henkin's Ford Bridge, Shoal Creek, Proctorville Vic., Caldwell County, Missouri, 1887. King Iron Bridge & Manufacturing Company (Cleveland, OH), builder. Phillip Geller, photographer, 1997. P&P,HAER,MO,13-PROV.V,1-9.

3-331. Detail of pin connection at bottom chord, Henkin's Ford Bridge, Shoal Creek, Proctorville Vic., Caldwell County, Missouri, 1887. King Iron & Manufacturing Bridge Company (Cleveland, OH), builder. Phillip Geller, photographer, 1997. P&P,HAER,MO, 13-PROV.V,1-8.

3-332

3-333. Finderne Avenue Bridge, railroad overpass, Bridgewater Township, Somerset County, New Jersey, 1913. Joseph Osgood, Central Railroad of New Jersey, engineer; Phoenix Bridge Company (Phoenixville, PA), fabricator. Ken DeBlieu, photographer, 1982. P&P,HAER,NJ,18-BRIWA,1-3.

3-332. Siuslaw River Bridge, Walton Vic., Lane County, Oregon, 1912 (moved to present site from Crooked River near Prineville 1956). Coast Bridge Company (Portland, OR), designer; Northwest Steel Company, fabricator. Jerry Robertson, photographer, 1986. P&P,HAER,ORE,20-WALT.V,1-3.

This 124-foot span is the oldest extant riveted through truss in Oregon. An example of an early-twentieth-century variant of the Warren truss that superimposes two triangular webs, it is known as a double-intersection or quadrangular truss.

3-334. West Wishkah Bridge, Wishkah River Middle Fork, Aberden Vic., Grays Harbor County, Washington, 1915. F. D. Sheffield (Seattle, WA), engineer; Coast Bridge Company (Portland, OR), builder. Steve Vento, photographer, 1988. P&P,HAER,WASH,14-ABER.V,1-3.

3-333

3-334

3-335. Wolf Creek Bridge, Missouri River, Craig Vic., Lewis and Clark County, Montana, 1932–1933. Benedikt J. Omburn, Montana Highway Commission, engineer; W. P. Roscoe Company (Billings, MT), builder; Minneapolis Steel & Machinery Company and Colorado Fuel & Iron Company (Pueblo, CO), fabricators. Jet T. Lowe, photographer, 1980. P&P,HAER,MONT,25-CRA.V,1-2.

A project commissioned by the state to relieve unemployment during the Great Depression, this continuous riveted Warren through truss has three spans of 135, 180, and 135 feet.

3-336. Dunnville Bridge, Red Cedar River, Downsville Vic., Dunn County, Wisconsin, 1934. State Highway Commission, designer; Worden-Allen Company (Milwaukee, WI), builder. Jerry Mathiason, photographer, 1996. P&P,HAER,WIS,17-DOW.V,1-1.

This riveted Warren through truss has verticals and additional bracing by diagonal and horizontal members.

3-337. Santa Fe Railroad Bridge, Illinois & Michigan Canal, Joliet, Will County, Illinois, ca. 1935. American Bridge Company (Pittsburgh, PA), builder. Martin Stupich, photographer, 1989. P&P,HAER,ILL,99-JOL,8-3.

This 200-foot skewed span is a riveted steel Warren truss with vertical and double-intersection diagonals.

3-335

3-336

3-337

Warren Pony Truss

3-338

3-338. Weston Road Bridge, spanning Conrail tracks, Wellesley, Norfolk County, Massachusetts, 1888. R. F. Hawkins Iron Works (Springfield, MA), builder. Martin Stupich, photographer, 1990. P&P,HAER,MASS,11-WEL,5-3.

The Boston & Albany Railroad built this 74-foot, riveted double-intersection Warren pony truss to eliminate a grade crossing.

3-339. Detail, Weston Road Bridge, spanning Conrail tracks, Wellesley, Norfolk County, Massachusetts, 1888. R. F. Hawkins Iron Works (Springfield, MA), builder. Martin Stupich, photographer, 1990. P&P,HAER,MASS,11-WEL,5-5.

3-340. Range Line Road Bridge, West Branch Eau Claire River, Langlade County, Wisconsin, 1908. Wisconsin Bridge & Iron Company (Milwaukee, WI), builder. Martin Stupich, photographer, 1987. P&P,HAER,WIS,34-ACK,1-1.

3-341. Section and details, Range Line Road Bridge, West Branch Eau Claire River, Langlade County, Wisconsin, 1908. Wisconsin Bridge & Iron Company (Milwaukee, WI), builder. J. Bilello, delineator, 1987. P&P,HAER,WIS,34-ACK,1-,sheet no. 2.

3-339

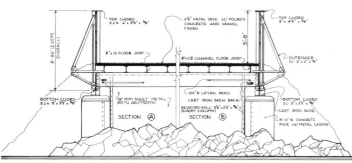

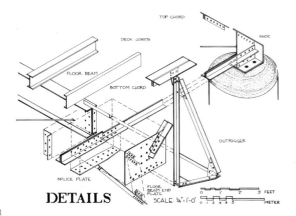

3-340

3-341

3-342. Smith Bridge, Meduxnekeag River, Houlton, Aroostook, Maine, 1910. Hardy S. Ferguson, designer; Frank Lowery, builder. Brian Vandenbrink, photographer, 1993. P&P,HAER,ME,2-HOUL,1-4.

3-343. Shell Creek Bridge, Shell Vic., Big Horn County, Wyoming, 1920. Midland Bridge Company (Kansas City, MO), builder. Clayton B. Fraser, photographer, 1992. P&P,HAER,WYO,2-SHEL.V,1-1.

3-344. Honeydew Creek Bridge, Honeydew, Humboldt County, California, 1925. Arthur Logan, county surveyor, designer; Frank Kelley, builder. Dave Swanlund, photographer, 1979. P&P,HAER,CAL,12-HOND,1-2.

The steel truss has been inverted.

3-342

3-343

3-344

3-345

3-346

3-347

3-345. Bridge 4759, Cannon River, Cannon Falls, Goodhue County, Minnesota, 1928. Minnesota Highway Department, designer; Guaranty Construction Company (Minneapolis, MN), builder. Jerry Mathiason, photographer, 1995. P&P,HAER,MINN,25-CANFA,1-3.

3-346. Underside of deck, Bridge 4759, Cannon River, Cannon Falls, Goodhue County, Minnesota, 1928. Minnesota Highway Department, designer; Guaranty Construction Company (Minneapolis, MN), builder. Jerry Mathiason, photographer, 1995. P&P,HAER,MINN,25-CANFA,1-7.

3-347. Thompson's Station Bridge, White Clay Creek, Newark, New Castle County, Delaware, 1928. Charles E. Grubb, county engineer, designer; Belmont Iron Works (Philadelphia, PA), fabricator. Tim O'Brian, photographer, 1993. P&P,HAER,DEL,2-NEWCA,47-4.

3-349

3-348

Warren Deck Truss

3-348. Smith Avenue High Bridge, Mississippi River, St. Paul, Minnesota, 1887–1889 (demolished 1985). Keystone Bridge Company (Pittsburgh, PA), builder. David R. Gonzalez, photographer, 1984. P&P,HAER,MINN,62-SAIPA,12-2.

3-349. Dead Indian Canyon Bridge, Grand Canyon National Park Vic., Coconino County, Arizona, 1933–1934. U.S. Bureau of Public Roads and National Park Service, designers; Vinson & Pringle (Phoenix, AZ), builder. Brian C. Grogan, photographer, 1994. P&P,HAER,ARIZ-3-GRACAN.V,1-3.

3-350. North Fork Bridge, White River, Norfolk, Baxter County, Arkansas, 1937. Arkansas State Highway Commission, designer; Vincennes Steel Corporation (Vincennes, IN), builder. Louise T. Cawood, photographer, 1988. P&P,HAER,ARK,3-NORF,1-4.

3-351. Riveted connections at bottom chord, North Fork Bridge, White River, Norfolk, Baxter County, Arkansas, 1937. Arkansas State Highway Commission, designer; Vincennes Steel Corporation (Vincennes, IN), builder. Louise T. Cawood, photographer, 1988. P&P,HAER,ARK,3-NORF,1-10.

3-350

3-351

3-352. Pit River Bridge, Southern Pacific Railroad, Redding Vic., Shasta County, California, 1941. Russell Lee, photographer, 1941. P&P,LC-USF34-071240-D.

3-353. Construction of bottom chords showing girders and ties, Pit River Bridge, Southern Pacific Railroad, Redding Vic., Shasta County, California, 1941. Russell Lee, photographer, 1941. P&P,LC-USF34-071108-D.

3-354. Construction workers at lunch, Pit River Bridge, Southern Pacific Railroad, Redding Vic., Shasta County, California, 1941. Russell Lee, photographer, 1941. P&P,LC-USF34-071170-D.

CANTILEVERED TRUSS

The falsework required to support the construction of arches and simple span trusses can be costly and difficult to erect, and its presence in navigable waterways can pose a hazard for shipping. Cantilevered trusses are largely self-supporting during construction and can be built without falsework. Although the principle of cantilevered construction was not unknown, few American builders considered applying it to bridges before the late 1860s. The success of Charles Shaler Smith's Kentucky River High Bridge (1877; 3-356) and long-span cantilevers in Europe spurred the use of the technique, except for a time following the 1907 collapse of the nearly completed Quebec Bridge crossing the St. Lawrence River. Variants of Pratt and Warren trusses are the most common configurations for cantilevered trusses.

3-355. "Champion's Double Lever Bridge," illustrated broadsheet by Samuel and Thomas Champion, Washington, D.C., 1853. RBSCD, Printed Ephemera Collection, Portfolio 201, Folder 13a.

In this proposal, the Champions applied the suspension principles of the Bollman truss to a bridge conceived as a pair of double cantilevers.

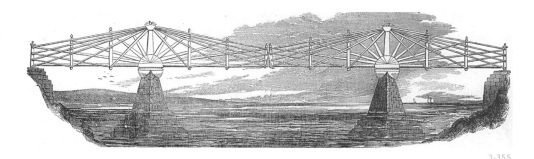

3-355

3-356. High Bridge, Kentucky River, High Bridge, Jessamine County, Kentucky, 1875–1877 (replaced 1910–1911). Charles Shaler Smith, designer. Unidentified photographer, ca. 1902. P&P, LC-USZ62-119125.

Charles Shaler Smith (1836–1883) was among the foremost engineers practicing in the United States in the decades following the Civil War. His engineering firm, Smith, Latrobe & Co., later known as the Baltimore Bridge Company, had a national practice designing and building bridges (3-307, 3-310) and other structures, primarily for railroads. Built for the Cincinnati Southern Railroad, the High Bridge was the first major cantilever bridge in North America. Smith cantilevered the Whipple-Murphy trusses to avoid the expense of erecting falsework in the deep gorge. The towers visible at the ends of the span had been built in the early 1850s by John Roebling for an unrealized suspension bridge. In 1911, Smith's bridge was replaced by the present structure designed by Gustav Lindenthal.

3-356

3-357

3-357. Michigan Central Railway Bridge (Great Cantilever Bridge), Niagara River, Niagara Falls, New York, 1883 (demolished 1925). Charles C. Schneider, chief engineer; Central Bridge Company (Buffalo, NY), builder. Engraving published in Scientific American 49, no. 22 (December 1, 1883), 335. P&P,LC-USZ62-61618.

Schneider spanned the 470 feet between the steel bents with two cantilevers composed of steel and wrought-iron Whipple-Murphy deck trusses that support a 120-foot Pratt truss inserted between them.

3-358. Poughkeepsie Railroad Bridge, Hudson River, Poughkeepsie, Dutchess County, New York, completed 1888. J. F. O'Rourke, P. P. Dickinson, engineers; Manhattan Bridge Company (Philadelphia, PA), contractor; Union Bridge Company (Athens, PA), superstructure erection. William Henry Jackson, photographer, between 1888 and 1897. P&P,LC-D4-4015.

The bridge has two 548-foot cantilevered deck trusses.

3-359. Central Bridge, Ohio River, Cincinnati, Ohio, and Newport, Kentucky, 1890–1891. L. P. G. Bouscaren, A. H. Porter, F. C. Osborn, engineers; King Iron Bridge & Manufacturing Company (Cleveland, OH), builders. J. M. Fiegel, J. L. Metille Jr., R. M. Morris, photographers, 1989. P&P,HAER,KY,19-NEWP,2-6.

This highway bridge consists of three through truss cantilevers, two Pennsylvania trusses, and a Pratt truss.

3-360. Kansas City, Fort Scott & Memphis Railroad Bridge, Mississippi River, Memphis, Tennessee, 1888–1892. George S. Morison and Alfred Noble, engineers. DETR, ca. 1906. P&P,LC-D4-19384.

The bridge has 2,597 feet of trussed spans, including a 791-foot cantilevered main channel span, the longest American railroad span of the nineteenth century.

3-358

3-359

3-360

04504 WINONA BRIDGES OVER THE MISSISSIPI.

COPYRIGHTED 1898 DETROIT PHOTOGRAPHIC CO

3-361

3-361. Mississippi River Wagon Bridge, Winona, Winona County, Minnesota, 1892 (demolished 1942). DETR, 1892. P&P,LC-D4-4504.

In the background is the railroad swing bridge built in 1890–1891 (see 4-008).

3-362. Northampton Street Bridge, Delaware River, Easton, Northampton County, Pennsylvania, 1895–1896. James Madison Porter III, engineer; Union Bridge Company (Athens, PA), builder; Belmont Iron Works (Philadelphia, PA), fabricator. Jet T. Lowe, photographer, 1999. P&P,HAER,PA,48-EATO,15-3.

The long eyebars forming the upper chords of the trusses, a technique also employed on the Queensboro Bridge (3-364–3-367), have a profile resembling a suspension bridge, but the two cantilevered arms are structurally independent, meeting at mid-span.

3-362

3-363

3-363. Pittsburgh & Lake Erie Railroad, Ohio River, Beaver, Beaver County, Pennsylvania, 1908–1910. A. R. Raymer, Albert Lucius, Paul L. Wolfel, engineers; McClintic-Marshall Construction Company (Pittsburgh, PA), superstructure; Dravo Contracting Company (Pittsburgh, PA), substructure. Jet T. Lowe, photographer, 1999. P&P,HAER,PA,4-BEAV,1-5.

3-364. Queensboro Bridge (Blackwell's Island Bridge), East River, New York, New York, 1901–1909. Gustav Lindenthal, chief engineer; Henry Hornbostel, architect; Pennsylvania Steel Company (Steelton, PA), builder. DETR, between 1910 and 1920. P&P,LC-D4-72133.

The program specifying four streetcar tracks, two elevated-railway tracks, a roadway, and footwalks required Lindenthal to design an extraordinarily heavy steel structure. Given the bridge's visual prominence, the city bridge department hired architect Henry Hornbostel as a consultant on aesthetic issues. Lindenthal and Hornbostel also worked together on the Hell Gate Bridge (see 2-072).

3-364

3-365. Detail of cantilevered truss, Queensboro Bridge (Black-well's Island Bridge), East River, New York, New York, 1901–1909. Gustav Lindenthal, chief engineer; Henry Hornbostel, architect; Pennsylvania Steel Company (Steelton, PA), builder. Jack E. Boucher, photographer, ca. 1970. P&P,HAER,NY,31-NEYO,160-7.

Unlike many cantilevered bridges, the Queensboro does not have a suspended span between the arms.

3-366. Steel workers on deck, Queensboro Bridge (Blackwell's Island Bridge), East River, New York, New York, 1901–1909. Gustav Lindenthal, chief engineer; Henry Hornbostel, architect; Pennsylvania Steel Company (Steelton, PA), builder. Unidentified photographer, 1907. P&P,LC-USZ62-98739.

3-367. Steel workers on top of tower, Queensboro Bridge (Blackwell's Island Bridge), East River, New York, New York, 1901–1909. Gustav Lindenthal, chief engineer; Henry Hornbostel, architect; Pennsylvania Steel Company (Steelton, PA), builder. Unidentified photographer, 1907. P&P,LC-USZ62-119527.

3-365

3-367

3-366

3-368

3-369

3-368. Pulaski Skyway, Hackensack River, Jersey City, New Jersey, 1924–1932. Sigvald Johannesson, engineer; New Jersey State Highway Deparment, designer. Jack E. Boucher, photographer, 1978. P&P,HAER,NJ,9-JERCI,10-5.

3-369. Smith River Bridge, Crescent City Vic., Del Norte County, California, 1928–1929 (demolished 1989). California Division of Highways, designer; Parker-Schram Company, builder; Virginia Bridge & Iron Works (Roanoke, VA), steel fabricator. Don Tateishi, photographer, 1989. P&P,HAER,CAL,8-CRECI.V,1-4.

3-370. Twin Falls–Jerome Bridge, Snake River, Twin Falls, Twin Falls County, Idaho, 1926–1927 (demolished ca. 1946). R. M. Murray, engineer; Union Bridge Company (Portland, OR), builder. Fred Powers and Clarence Bisbee, photographers, 1927. P&P,HAER,ID,42-TWIFA,1-11.

3-371. Construction of the suspended truss between cantilevered arms, Twin Falls–Jerome Bridge, Snake River, Twin Falls, Twin Falls County, Idaho, 1926–1927 (demolished ca. 1946). R. M. Murray, engineer; Union Bridge Company (Portland, OR), builder. Fred Powers and Clarence Bisbee, photographers, 1927. P&P,HAER,ID,42-TWIFA,1-8.

3-372. Surveyor, Twin Falls–Jerome Bridge, Snake River, Twin Falls, Twin Falls County, Idaho, 1926–1927 (demolished ca. 1946). R. M. Murray, engineer; Union Bridge Company (Portland, OR), builder. Fred Powers and Clarence Bisbee, photographers, 1927. P&P,HAER,ID,42-TWIFA,1-9.

3-370

3-371

3-372

3-373

3-374

3-373. Longview Bridge, Columbia River, Longview, Cowlitz County, Washington, 1930. Joseph B. Strauss, chief engineer; Bethlehem Steel Company (Bethlehem, PA), general contractor. Jet T. Lowe, photographer, 1993. P&P,HAER,WASH,8-LONVI,1-3.

The central span is 1,200 feet.

3-374. South tower of main cantilever, Longview Bridge, Columbia River, Longview, Cowlitz County, Washington, 1930. Joseph B. Strauss, chief engineer; Bethlehem Steel Company (Bethlehem, PA), general contractor. Jet T. Lowe, photographer, 1993. P&P,HAER,WASH,8-LONVI,1-5.

3-375

3-376

3-375. Clarendon Bridge, White River, Clarendon, Monroe County, Arkansas, 1931. Ira G. Hedrick, engineer; Austin Bridge Company (Dallas, TX), builder. Louise T. Cawood, photographer, 1988. P&P,HAER,ARK,48-CLAR,2-6.

The 400-foot cantilever consisting of two 120-foot arms and a 160-foot Warren-truss suspended span between them is one of three cantilevered bridges of similar dimensions Hedrick built across the White River around 1930. The use of cantilevers eliminated the need for falsework, which would have been at risk during the river's often violent floods.

3-376. Augusta Bridge, White River, Augusta, Woodruff County, Arkansas, 1929–1930 (demolished 2001). Ira G. Hedrick, engineer; Missouri Valley Bridge & Iron Company (Leavenworth, KS), builder. Louise T. Cawood, photographer, 1983. P&P,HAER,ARK,74-AUG,1-3.

3-377. Newport Bridge, White River, Newport, Jackson County, Arkansas, 1929–1930. Ira G. Hedrick, designer; Missouri Valley Bridge & Iron Company (Leavenworth, KS), builder. Louise T. Cawood, photographer, 1983. P&P,HAER,ARK,34-NEPO,1-4.

3-378. West pier, Newport Bridge, White River, Newport, Jackson County, Arkansas, 1929–1930. Ira G. Hedrick, engineer; Missouri Valley Bridge & Iron Company (Leavenworth, KS), builder. Louise T. Cawood, photographer, 1983. P&P,HAER,ARK,34-NEPO,1-6.

3-377

3-378

3-379

3-380

3-381

3-379. East Bay crossing, San Francisco–Oakland Bay Bridge, San Francisco Bay, between San Francisco and Oakland, California, 1933–1936. C. H. Purcell, chief engineer; Bridge Builders Inc., substructure; Columbia Steel Company (Portland, OH), and American Bridge Company (Pittsburgh, PA), superstructure. Jet T. Lowe, photographer, 1985. P&P,HAER,CAL,38-SANFRA,141-8.

The San Francisco Bay crossing consists of two bridges and a tunnel through Yerba Buena Island. The western span is a suspension bridge (see 5-093), and the eastern bridge is composed of trusses including a 1,400-foot cantilevered span. Construction of a new bridge less susceptible to earthquakes began in 2002.

3-380. Coos Bay Bridge (Conde B. McCullough Memorial Bridge), North Bend, Coos County, Oregon, 1934–1936. Conde B. McCullough, engineer; Virginia Bridge & Iron Company (Roanoke, VA), superstructure; Northwest Roads Company, concrete. Jet T. Lowe, photographer, 1990. P&P,HAER,ORE,6-NOBE,1-11.

The busy shipping channel precluded the use of falsework, so McCullough spanned it with a steel cantilevered truss. The remainder of the bridge consists of reinforced concrete deck arches and deck girder approach spans. The construction of the arches represents an early adoption in the United States of the Considère hinge (see 2-173), a technique that allowed the connections of the crowns and skewbacks to be completed after the dead load was placed on the arch. Always attentive to the aesthetic character of his bridges, McCullough shaped the chords and sway braces of the cantilever to complement the rhythm of the arches.

3-381. Interior view of south portal, Coos Bay Bridge (Conde B. McCullough Memorial Bridge), North Bend, Coos County, Oregon, 1934–1936. Conde B. McCullough, engineer; Virginia Bridge & Iron Company (Roanoke, VA), superstructure; Northwest Roads Company, concrete. Jet T. Lowe, photographer, 1990. P&P,HAER,ORE,6-NOBE,1-6.

3-382. View of piers supporting south approach span, Coos Bay Bridge (Conde B. McCullough Memorial Bridge), North Bend, Coos County, Oregon, 1934–1936. Conde B. McCullough, engineer; Virginia Bridge & Iron Company (Roanoke, VA), superstructure; Northwest Roads Company, concrete. Jet T. Lowe, photographer, 1990. P&P,HAER,ORE,6-NOBE,1-14.

3-383. Stairway at south abutment, Coos Bay Bridge (Conde B. McCullough Memorial Bridge), North Bend, Coos County, Oregon, 1934–1936. Conde B. McCullough, engineer; Virginia Bridge & Iron Company (Roanoke, VA), superstructure; Northwest Roads Company, concrete. Jet T. Lowe, photographer, 1990. P&P,HAER,ORE,6-NOBE,1-15.

McCullough envisioned his bridges as destinations as well as conveyances. At each end of the Coos Bay Bridge, he provided parking and elegant staircases leading to picnic areas.

3-382

3-383

3-384

3-385

3-386

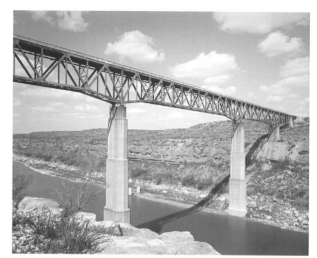
3-387

3-384. Blue Water Bridge, St. Clair River, Port Huron, St. Clair County, Michigan, and Point Edward, Ontario, Canada, 1936–1938. Modjeski & Masters (Harrisburg, PA), engineers; Paul Cret, consulting architect; American Bridge Company (Pittsburgh, PA), builder. Clayton B. Fraser, photographer, 1994. P&P,HAER,MICH,74-POHU,1A-12.

The cantilevered main span of this Warren through truss is 871 feet.

3-385. Toll plaza, Blue Water Bridge, St. Clair River, Port Huron, St. Clair County, Michigan, and Point Edward, Ontario, Canada, 1936–1938. Paul Cret, consulting architect. Clayton B. Fraser, photographer, 1994. P&P,HAER,MICH,74-POHU,1-14.

3-386. Toll plaza and administration building, Blue Water Bridge, St. Clair River, Port Huron, St. Clair County, Michigan, and Point Edward, Ontario, Canada, 1936–1938. Paul Cret, consulting architect. Clayton B. Fraser, photographer, 1994. P&P,HAER, MICH,74-POHU,1-18.

3-387. Pecos River High Bridge, Southern Pacific Railroad, Langtry Vic., Val Verde County, Texas, 1943–1944. Modjeski & Masters (Harrisburg, PA), engineers; Bethlehem Steel Company (Chicago, IL), superstructure; Brown & Root (Houston, TX), substructure. Joseph Elliott, photographer, 1998. P&P,HAER,TX,233-LANG.V, 1-4.

This 1,390-foot continuous steel cantilever truss replaced the trestle built in 1892 (see 1-031).

3-388

3-388. Chicago Skyway (Calumet Skyway Toll Bridge), Calumet River, Chicago, Illinois, 1956–1958. J. E. Greimer Company, engineers; U.S. Steel (Pittsburgh, PA), superstructure; E. J. Albrecht Company, substructure. Jet T. Lowe, photographer, 1999. P&P,HAER,ILL,16-CHIG,138-18.

The 650-foot cantilevered Warren through truss is the centerpiece of a nearly 8-mile highway that links Chicago to the Indiana Toll Road.

3-389. Toll plaza, Chicago Skyway (Calumet Skyway Toll Bridge), Calumet River, Chicago, Illinois, 1956–1958. Consoer, Townsend & Associates (Chicago, IL), designer. Jet T. Lowe, photographer, 1999. P&P,HAER,ILL,16-CHIG,138A-5.

3-390. Service building beneath toll plaza, Chicago Skyway (Calumet Skyway Toll Bridge), Calumet River, Chicago, Illinois, 1956–1958. Consoer, Townsend & Associates (Chicago, IL), designer. Jet T. Lowe, photographer, 1999. P&P,HAER,ILL,16-CHIG,138A-6.

3-389

3-390

MOVABLE BRIDGES

Movable spans that can be raised or pivoted are used for bridges crossing navigable waterways when topographic or economic conditions do not permit the construction of fixed spans above the height of ship traffic. There are three basic types of movable spans: swing bridges that rotate horizontally, bascule bridges (popularly known as drawbridges) that tilt up, and lift bridges that hoist the span between a pair of towers. Challenges in the design of all three arise in balancing components and engineering the pivot points to minimize the force required to move the span with a high degree of control. Movable bridges are rarely built today, because their operation interrupts the smooth flow of traffic and incurs expenses for the maintenance of machinery and the salaries of operators.

Swing bridges may be built with the pivot either at the center or at one end of the movable span (sometimes referred to as a bob-tail swing span). Center pivots allow longer spans than end pivots but require mid-channel piers that can pose an obstacle for navigation. End pivots, however, occupy space along the riverbank for which there may be competing uses.

4-001

4-001. "View of Rush Street Bridge & c. from Nortons Block River Street," Chicago River, Chicago, Illinois. Color lithograph, E. White-field, delineator; Charles Shober (Chicago, IL), lithographer and publisher, ca. 1861. P&P,LC-USZC4-2334.

Between 1860 and 1890, the City of Chicago built fifty-five center-pivot bridges across its inland waterways. The Rush Street Bridge was the first on the Chicago River. The relatively light weight of the bowstring truss made it a popular structural solution for the movable spans.

4-002. Lake Street Bridge, Chicago River, Chicago, Illinois. Lithograph in James W. Sheahan, Chicago Illustrated (Chicago: Jevne & Almini, 1866). P&P,LC-USZ62-23805.

This span consisted of an arched lattice truss.

4-002

4-003. Detail of site plan, Missouri River Railroad Bridge (Wabash Railroad) project, St. Charles, St. Charles County, Missouri, 1865. Octave Chanute, engineer. Manuscripts, Octave Chanute collection, oversize box, unnumbered sheet.

Octave Chanute (1832–1910) learned engineering from the ground up, beginning his career as a chainman assisting railroad surveyors and attaining the position of chief engineer for the Chicago & Alton Railroad fifteen years later. In 1865, he responded to a call for proposals for a railroad bridge crossing the Missouri River at St. Charles. In this drawing, he criticized a proposed location, noting that the current deflected by the sandbar would make contact with the piers at an undesirable oblique angle. He proposed a site slightly to the south where the current ran parallel to the east and west banks and a more orthogonal alignment of the piers could be achieved.

4-004. Detail, Missouri River Railroad Bridge (Wabash Railroad) project, St. Charles, St. Charles County, Missouri, 1865. Octave Chanute, engineer. Manuscripts, Octave Chanute collection, oversize box, sheet no. 5.

Chanute proposed spanning the river with eight 190-foot Howe through trusses and a 295-foot Howe truss swing span. Given the high price of iron at the end of the Civil War, he recommended that the bridge be built of wood and that the money saved be invested to provide a funding source that would pay for its periodic replacement.

4-005. Detail of protective siding, Missouri River Railroad Bridge (Wabash Railroad) project, St. Charles, St. Charles County, Missouri, 1865. Octave Chanute, engineer. Manuscripts, Octave Chanute collection, oversize box, sheet no. 7.

Chanute argued that cladding the structural members with wooden siding rather than enclosing the entire structure as a covered bridge would be more economical and would reduce the exposure to and resultant stresses from high winds. It also would offer protection from the corrosive exhaust of locomotives.

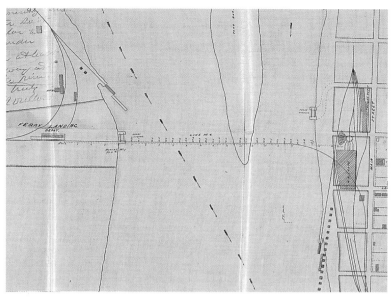

4-003

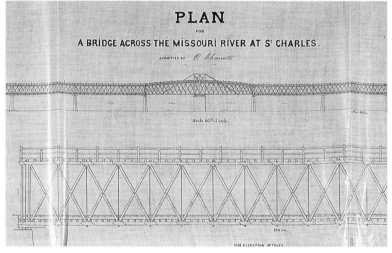

4-004

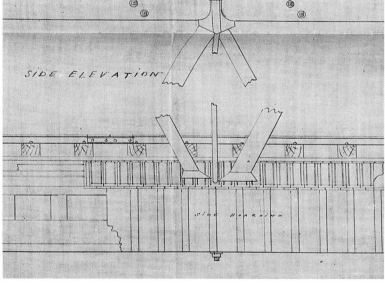

4-005

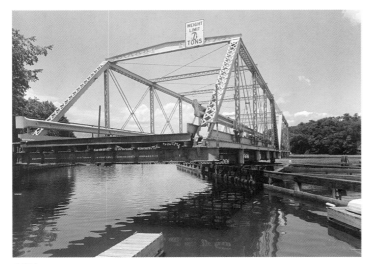
4-006

4-007

4-006. Saugatuck River Bridge, Westport, Fairfield County, Connecticut, 1884. Central Bridge Company (Buffalo, NY), builder. Robert Moore, photographer, 1990. P&P,HAER,CONN,1-WESPO,15-13.

This 142-foot center-pivot swing span is a rare surviving example of a manually operated swing bridge.

4-007. Workers manually operating pivot mechanism, Saugatuck River Bridge, Westport, Fairfield County, Connecticut, 1884. Central Bridge Company (Buffalo, NY), builder. Robert Moore, photographer, 1990. P&P,HAER,CONN,1-WESPO,15-14.

4-008. Winona Railroad Bridge, Mississippi River, Winona, Winona County, Minnesota, 1890–1891 (demolished 1990). George S. Morison, engineer; Union Bridge Company (Athens, PA), builder. Clayton B. Fraser, photographer, 1985. P&P,HAER,MINN,85-WIN,1-9.

The 420-foot swing span is a modified Pratt truss composed of steel and wrought-iron members. The drum supporting the span revolves on an iron track and was originally powered by a steam engine.

4-008

4-009. Swing span, Winona Railroad Bridge, Mississippi River, Winona, Winona County, Minnesota, 1890–1891 (demolished 1990). George S. Morison, engineer; Union Bridge Company (Athens, PA), builder. Clayton B. Fraser, photographer, 1985. P&P,HAER,MINN,85-WIN,1-31.

4-010. Railroad track coupling at swing span, Winona Railroad Bridge, Mississippi River, Winona, Winona County, Minnesota, 1890–1891 (demolished 1990). George S. Morison, engineer; Union Bridge Company (Athens, PA), builder. Clayton B. Fraser, photographer, 1985. P&P,HAER,MINN,85-WIN, 1-46.

4-011. Interior of bridge tender's house, Winona Railroad Bridge, Mississippi River, Winona, Winona County, Minnesota, 1890–1891 (demolished 1990). George S. Morison, engineer; Union Bridge Company (Athens, PA), builder. Clayton B. Fraser, photographer, 1985. P&P,HAER,MINN,85-WIN,1-36.

4-009

4-011

4-010

4-012. Rim-bearing turntable with fender composed of timber framing and pilings, Winona Railroad Bridge, Mississippi River, Winona, Winona County, Minnesota, 1890–1891 (demolished 1990). George S. Morison, engineer; Union Bridge Company (Athens, PA), builder. Clayton B. Fraser, photographer, 1985. P&P,HAER,MINN,85-WIN,1-38.

4-013. Rim-bearing turntable drum, wheels, and bull gear, Winona Railroad Bridge, Mississippi River, Winona, Winona County, Minnesota, 1890–1891 (demolished 1990). George S. Morison, engineer; Union Bridge Company (Athens, PA), builder. Clayton B. Fraser, photographer, 1985. P&P,HAER,MINN,85-WIN,1-41.

4-012

4-013

4-014

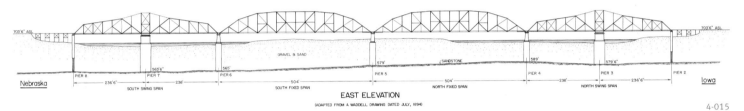

703'6" ASL

Nebraska

PIER 8   236'6"   PIER 7   238'   PIER 6   504'   PIER 5   504'   PIER 4   238'   PIER 3   236'6"   PIER 2

563'6"   565'   579'   589'   579'6"

GRAVEL & SAND   SANDSTONE

SOUTH SWING SPAN   SOUTH FIXED SPAN   NORTH FIXED SPAN   NORTH SWING SPAN

703'6" ASL

Iowa

EAST ELEVATION

(ADAPTED FROM A WADDELL DRAWING DATED JULY, 1894)

4-015

4-014. Pacific Short Line Bridge, Missouri River, Sioux City, Woodbury County, Iowa, 1890–1896 (demolished 1980). J. A. L. Waddell, chief engineer; Charles Sooy Smith, engineer, substructure; Phoenix Bridge Company (Phoenixville, PA), builder. Robert Ryan, Hans Muessig, photographers, 1980. P&P,HAER,IOWA,97-SIOCI,1-8.

4-015. Elevation, Pacific Short Line Bridge, Missouri River, Sioux City, Woodbury County, Iowa, 1890–1896 (demolished 1980). J. A. L. Waddell, chief engineer; Charles Sooy Smith, engineer, substructure; Phoenix Bridge Company (Phoenixville, PA), builder. Marie Neubauer, Hans Muessig, delineators, 1981. P&P,HAER,IOWA,97-SIOCI,1-,sheet no. 1.

4-016. Elevation and plan of swing span, Pacific Short Line Bridge, Missouri River, Sioux City, Woodbury County, Iowa, 1890–1896 (demolished 1980). J. A. L. Waddell, chief engineer; Charles Sooy Smith, engineer, substructure; Phoenix Bridge Company (Phoenixville, PA), builder. Marie Neubauer, Hans Muessig, delineators, 1981. P&P,HAER,IOWA,97-SIOCI,1-,sheet no. 2.

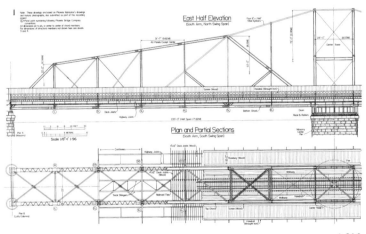

4-016

4-017

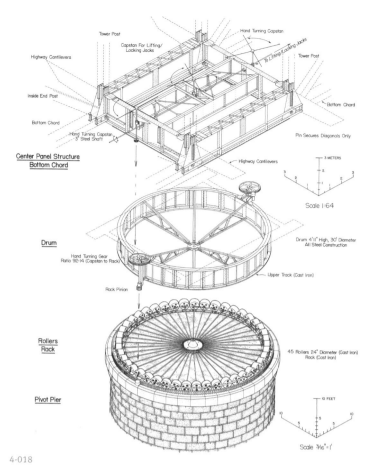

4-018

4-019

4-017. Pier, drum, and rollers of rim-bearing swing mechanism, south pivot, Pacific Short Line Bridge, Missouri River, Sioux City, Woodbury County, Iowa, 1890–1896 (demolished 1980). J. A. L. Waddell, chief engineer; Charles Sooy Smith, engineer, substructure; Phoenix Bridge Company (Phoenixville, PA), builder. Robert Ryan, Hans Muessig, photographers, 1980. P&P,HAER,IOWA, 97-SIOCI,1-29.

4-018. Details of rim-bearing pivot mechanism, Pacific Short Line Bridge, Missouri River, Sioux City, Woodbury County, Iowa, 1890–1896 (demolished 1980). J. A. L. Waddell, chief engineer; Charles Sooy Smith, engineer, substructure; Phoenix Bridge Company (Phoenixville, PA), builder. Marie Neubauer, Hans Muessig, Richard K. Anderson Jr., delineators, 1981. P&P,HAER,IOWA,97-SIOCI,1-,sheet no. 5.

4-019. Drum turning mechanism, Pacific Short Line Bridge, Missouri River, Sioux City, Woodbury County, Iowa, 1890–1896 (demolished 1980). J. A. L. Waddell, chief engineer; Charles Sooy Smith, engineer, substructure; Phoenix Bridge Company (Phoenixville, PA), builder. Robert Ryan, Hans Moessig, photographers, 1980. P&P,HAER,IOWA,97-SIOCI,1-32.

4-020. Drum interior, Pacific Short Line Bridge, Missouri River, Sioux City, Woodbury County, Iowa, 1890–1896 (demolished 1980). J. A. L. Waddell, chief engineer; Charles Sooy Smith, engineer, substructure; Phoenix Bridge Company (Phoenixville, PA), builder. Robert Ryan, Hans Moessig, photographers, 1980. P&P,HAER,IOWA,97-SIOCI,1-31.

4-021. Locking mechanism, Pacific Short Line Bridge, Missouri River, Sioux City, Woodbury County, Iowa, 1890–1896 (demolished 1980). J. A. L. Waddell, chief engineer; Charles Sooy Smith, engineer, substructure; Phoenix Bridge Company (Phoenixville, PA), builder. Robert Ryan, Hans Muessig, photographers, 1980. P&P,HAER,IOWA,97-SIOCI,1-34.

4-022

4-023

4-024

4-022. Powow River Bridge, Amesbury, Essex County, Massachusetts, 1891. Boston Bridge Works, builder. Martin Stupich, photographer, 1990. P&P,HAER,MASS,5-AMB,4-6.

4-023. Bridgeport swing span bridge, Nashville, Chattanooga & St. Louis Railroad, Tennessee River, Bridgeport Vic., Jackson County, Alabama, 1890–1892 (demolished). Louisville Bridge & Iron Company, builder. C. N. Beasley, photographer, 1980. P&P,HAER,ALA,36-BRIPO.V,1-2.

4-024. Swing span in open position, Bridgeport swing span bridge, Nashville, Chattanooga & St. Louis Railroad, Tennessee River, Bridgeport Vic., Jackson County, Alabama, 1890–1892 (demolished). Louisville Bridge & Iron Company, builder. C. N. Beasley, photographer, 1980. P&P,HAER,ALA,36-BRIPO.V,1-13.

4-025

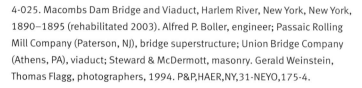

4-027

4-025. Macombs Dam Bridge and Viaduct, Harlem River, New York, New York, 1890–1895 (rehabilitated 2003). Alfred P. Boller, engineer; Passaic Rolling Mill Company (Paterson, NJ), bridge superstructure; Union Bridge Company (Athens, PA), viaduct; Steward & McDermott, masonry. Gerald Weinstein, Thomas Flagg, photographers, 1994. P&P,HAER,NY,31-NEYO,175-4.

4-026. Detail of rim-bearing swing span showing center of truss over pivot, Macombs Dam Bridge and Viaduct, Harlem River, New York, New York, 1890–1895 (rehabilitated 2003). Alfred P. Boller, engineer; Passaic Rolling Mill Company (Paterson, NJ), bridge superstructure; Union Bridge Company (Athens, PA), viaduct; Steward & McDermott, masonry. Gerald Weinstein, Thomas Flagg, photographers, 1994. P&P,HAER,NY,31-NEYO,175-31.

4-027. 155th Street Viaduct, Macombs Dam Bridge and Viaduct, Harlem River, New York, New York, 1890–1895 (rehabilitated 2003). Alfred P. Boller, engineer; Passaic Rolling Mill Company (Paterson, NJ), bridge superstructure; Union Bridge Company (Athens, PA), viaduct; Steward & McDermott, masonry. Gerald Weinstein, Thomas Flagg, photographers, 1994. P&P,HAER,NY,31-NEYO,175-2.

4-028. Approach from viaduct, Macombs Dam Bridge and Viaduct, Harlem River, New York, New York, 1890–1895 (rehabilitated 2003). Alfred P. Boller, engineer; Passaic Rolling Mill Company (Paterson, NJ), bridge superstructure; Union Bridge Company (Athens, PA), viaduct; Steward & McDermott, masonry. Gerald Weinstein, Thomas Flagg, photographers, 1994. P&P,HAER,NY,31-NEYO,175-25.

4-028

4-029

4-030

4-031

4-032

4-029. Rock Island Bridge (Government Bridge), Mississippi River, Rock Island, Rock Island County, Illinois, 1895–1896. Ralph Modjeski, engineer; Phoenix Bridge Company (Phoenixville, PA), builder. J Ceronie, photographer, 1985. P&P,HAER,ILL,81-ROCIL,3A-1.

4-030. Jackson Street Bridge, Passaic River, Newark, Essex County, New Jersey, 1897–1898. James Owen and Thomas McBarn, Essex and Hudson County engineering departments, designers; Fagan Iron Works (Jersey City, NJ), builder; Passaic Rolling Mill, fabricator. Gerald Weinstein, photographer, 1985. P&P,HAER,NJ,7-NEARK,18-3.

The riveted steel trusses of both the fixed and swing spans have an unusual configuration of web members and curved upper chords. Instead of conventional vertical and diagonal web members that do not intersect, the webs are composed entirely of diagonal members, which intersect with each other multiple times.

4-031. Interior view of swing span showing control house, Jackson Street Bridge, Passaic River, Newark, Essex County, New Jersey, 1897–1898. James Owen, Thomas McBarn, Essex and Hudson County engineering departments, designers; Fagan Iron Works (Jersey City, NJ), builder; Passaic Rolling Mill, fabricator. Gerald Weinstein, photographer, 1985. P&P,HAER,NJ,7-NEARK,18-8.

4-032. View of control house showing boilers and steam engine powering pivot mechanism, Jackson Street Bridge, Passaic River, Newark, Essex County, New Jersey, 1897–1898. James Owen, Thomas McBarn, Essex and Hudson County engineering departments, designers; Fagan Iron Works (Jersey City, NJ), builder; Passaic Rolling Mill, fabricator. Gerald Weinstein, photographer, 1985. P&P,HAER,NJ,7-NEARK,18-18.

4-033. Aerial view, Chicago, Milwaukee & St. Paul Railway, Bridge No. Z-6 (at bottom of photo), North Branch of Chicago River, Chicago, Illinois, 1898–1899. Onward Bates, engineer; American Bridge Works (Chicago, IL), fabricator. Jet T. Lowe, photographer, 1999. P&P,HAER,ILL,16-CHIG,120-1.

The pivot point of this 175-foot, plate-girder bob-tail swing bridge is located on land to maximize the navigable channel through this narrow branch of the Chicago River.

4-034. Built-up plate girders, Chicago, Milwaukee & St. Paul Railway, Bridge No. Z-6, North Branch of Chicago River, Chicago, Illinois, 1898–1899. Onward Bates, engineer; American Bridge Works (Chicago, IL), fabricator. Jet T. Lowe, photographer, 1999. P&P,HAER,ILL,16-CHIG,120-15.

4-035. Plate girder swing span, pivot rack, and rollers, Chicago, Milwaukee & St. Paul Railway, Bridge No. Z-6, North Branch of Chicago River, Chicago, Illinois, 1898–1899. Onward Bates, engineer; American Bridge Works (Chicago, IL), fabricator. Jet T. Lowe, photographer, 1999. P&P,HAER,ILL,16-CHIG, 120-11.

4-033

4-034

4-035

4-036

4-036. Burlington Northern Railroad Bridge, Willamette River, Portland, Oregon, 1906–1908. Ralph Modjeski, engineer. Alfred Staehli, photographer, 1985. P&P,HAER,ORE,26-PORT,8-41.

4-037. Center Street Bridge, Cuyahoga River, Cleveland, Ohio, 1900–1901. James Ritchie, James T. Pardee, engineers; King Bridge Company (Cleveland, OH), builder. Jet T. Lowe, photographer, 1978. P&P,HAER,OHIO,18-CLEV,29-4.

4-037

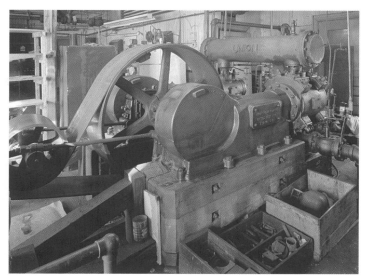

4-039

4-040

4-038. Northern Avenue Bridge, Fort Point Channel, Boston, Massachusetts, 1908. William Jackson, city engineer, designer. Jet T. Lowe, photographer, 1982. P&P,HAER,MASS,13-BOST,84-12.

The 283-foot-long swing span is 80 feet wide, allowing it to carry sidewalks, multiple traffic lanes, and railroad tracks. It rests on a 40-foot-diameter drum revolving on steel rollers driven by compressed air.

4-039. Compressed air machinery, Northern Avenue Bridge, Fort Point Channel, Boston, Massachusetts, 1908. William Jackson, city engineer, designer. Jet T. Lowe, photographer, 1982. P&P,HAER,MASS,13-BOST,84-30.

4-040. Interior of drum showing drive gear and rim-bearing rollers, Northern Avenue Bridge, Fort Point Channel, Boston, Massachusetts, 1908. William Jackson, city engineer, designer. Jet T. Lowe, photographer, 1982. P&P,HAER,MASS,13-BOST,84-36.

4-041

4-042

4-041. Gianella Bridge, Sacramento River, Hamilton City Vic., Glenn County, California, 1908–1911. John B. Leonard, engineer; Cotton Brothers (Oakland, CA), builder. Pete Asano, photographer, 1986. P&P,HAER,CAL,11-HAMCI.V,1-6.

4-042. Pier and rim-bearing pivot, Gianella Bridge, Sacramento River, Hamilton City Vic., Glenn County, California, 1908–1911. John B. Leonard, engineer; Cotton Brothers (Oakland, CA), builder. Pete Asano, photographer, 1986. P&P,HAER,CAL,11-HAMCI.V,1-11.

4-043

4-043. Moser Channel swing span, Seven Mile Bridge, Knight Key Vic., Monroe County, Florida, 1909–1912 (demolished 1981). Joseph C. Meredith, William J. Krome, chief engineers. Unidentified photographer, 1979. P&P,HAER,FLA,44-KNIKE,1-3.

True to its name, the Seven Mile Bridge consisted of a series of viaducts and bridges 35,600 feet in length built by the Florida East Coast Railway as part of its Key West Extension. It was converted for highway use in 1937–1938. The 253-foot swing span accommodated a shipping lane between the Gulf of Mexico and the Atlantic Ocean. A new bridge alongside it was completed in 1982.

4-044. Moser Channel swing span in 3/4 open position, Seven Mile Bridge, Knight Key Vic., Monroe County, Florida, 1909–1912 (demolished 1981). Joseph C. Meredith, William J. Krome, chief engineers. Unidentified photographer, 1979. P&P, HAER,FLA,44-KNIKE,1-4.

4-044

4-045

4-046

4-045. Fort Madison Bridge, Atchison, Topeka & Santa Fe Railroad, Mississippi River, Fort Madison, Lee County, Iowa, 1927. A. F. Robinson, Santa Fe Railroad, engineer; American Bridge Company (Pittsburgh, PA), superstructure; Union Bridge & Construction Company (Kansas City, MO), substructure. Joseph Elliott, photographer, 1995. P&P,HAER,IOWA,56-FMAD.1-5.

The bridge has double tracks on the lower deck and a two-lane highway on the upper deck. The main channel crossing consists of four fixed 270-foot through trusses and the swing span, which has two 266-foot arms.

4-046. Lower deck, Fort Madison Bridge, Atchison, Topeka & Santa Fe Railroad, Mississippi River, Fort Madison, Lee County, Iowa, 1927. A. F. Robinson, Santa Fe Railroad, engineer; American Bridge Company (Pittsburgh, PA), superstructure; Union Bridge & Construction Company (Kansas City, MO), substructure. Joseph Elliott, photographer, 1995. P&P,HAER,IOWA,56-FMAD.1-7.

4-047

4-047. Detail of swing span, Fort Madison Bridge, Atchison, Topeka & Santa Fe Railroad, Mississippi River, Fort Madison, Lee County, Iowa, 1927. A. F. Robinson, Santa Fe Railroad, engineer; American Bridge Company (Pittsburgh, PA), superstructure; Union Bridge & Construction Company (Kansas City, MO), substructure. Joseph Elliott, photographer, 1995. P&P,HAER,IOWA,56-FMAD.1-8.

4-048. Control room, Fort Madison Bridge, Atchison, Topeka & Santa Fe Railroad, Mississippi River, Fort Madison, Lee County, Iowa, 1927. A. F. Robinson, Santa Fe Railroad, engineer; American Bridge Company (Pittsburgh, PA), superstructure; Union Bridge & Construction Company (Kansas City, MO), substructure. Joseph Elliott, photographer, 1995. P&P,HAER,IOWA,56-FMAD.1-14.

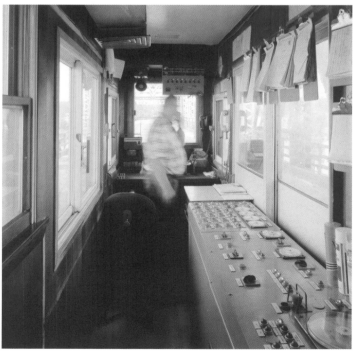

4-048

4-049

4-050

4-049. Bayou Teche Bridge, Ruth, St. Martin Parish, Louisiana, 1932. Louisiana Highway Commission, designer; Highway Maintenance Department, builder. Larry Norman, photographer, 1985. P&P,HAER,LA,50-RUTH,1-2.

The 106-foot manually driven bridge composed of steel I-beams has been strengthened by the addition of a tower and cables.

4-050. Workers turning capstan (winch) to close bridge, Bayou Teche Bridge, Ruth, St. Martin Parish, Louisiana, 1932. Louisiana Highway Commission, designer; Highway Maintenance Department, builder. Larry Norman, photographer, 1985. P&P,HAER,LA, 50-RUTH,1-5.

4-051. Umpqua River Bridge, Reedsport, Douglas County, Oregon, 1934–1936. Conde B. McCullough, engineer; Teufel & Carlson (Seattle, WA), builder. Jet T. Lowe, photographer, 1990. P&P,HAER,ORE,10-REPO,1-1.

Part of the Oregon Coast Highway, this bridge has a steel swing span with a curved upper chord that complements the shape of the reinforced concrete, tied-arch spans flanking it.

4-052. Swing span sway bracing details, Umpqua River Bridge, Reedsport, Douglas County, Oregon, 1934–1936. Conde B. McCullough, engineer; Teufel & Carlson (Seattle, WA), builder. Jet T. Lowe, photographer, 1990. P&P,HAER,ORE,10-REPO,1-11.

4-053

4-054

4-055

4-053. Swing span, Umpqua River Bridge, Reedsport, Douglas County, Oregon, 1934–1936. Conde B. McCullough, engineer; Teufel & Carlson (Seattle, WA), builder. Jet T. Lowe, photographer, 1990. P&P,HAER,ORE,10-REPO,1-9.

The swing span is a 430-foot, steel Parker through truss.

4-054. Turntable and drive motor, Umpqua River Bridge, Reedsport, Douglas County, Oregon, 1934–1936. Conde B. McCullough, engineer; Teufel & Carlson (Seattle, WA), builder. Jet T. Lowe, photographer, 1990. P&P,HAER,ORE,10-REPO,1-14.

The turntable mechanism is powered by a 60-horsepower electric motor.

4-055. George P. Coleman Memorial Bridge, York River, Yorktown, York County, Virginia, 1950–1952 (widened and reconstructed 1996). Parsons, Brinkerhoff, Hall & Macdonald (New York, NY), designers; American Bridge Company (Roanoke, VA), fabricator; Massman Construction Company, Kansas City Bridge Company, builders. Rob Tucher, photographer, 1993. P&P,HAER,VA,100-YORK,19-22.

The heavily used shipping lanes required two paired 500-foot swing spans.

BASCULE BRIDGES

Bascule bridges, which take their name from the French word for seesaw, consist of one or more counterweighted leaves that can be rotated upward about a horizontal pivot. The modern bascule bridge was developed around the turn of the twentieth century, drawing on the availability of structural steel for the trusses and connections and reliable electric motors to power the lift machinery. A number of engineers including Joseph Strauss, Alexander von Babo, Theodore Rall, and William Scherzer patented designs for pivot mechanisms and arrangements of counterweights above or below the deck. Compact and offering efficient, rapid operation, bascule bridges rapidly replaced swing spans, especially along the canals and narrow rivers winding through cities where space along the banks was limited and mid-channel piers would impede navigation.

Bascule Bridges in Chicago

By the 1890s, the mid-channel piers of the center-pivot bridges had become impediments to the passage of the larger vessels plying Chicago's narrow inland waterways. On land, people navigating the city's crowded streets were impatient with the length of time it took to open and close the spans. Under the direction of John Ericson, the Chicago Department of Public Works made the city a laboratory for the design of movable bridges during the first decades of the twentieth century and developed the Chicago-type bascule, which engineers throughout the country subsequently adopted. In addition to the Department of Public Works, other public agencies and private entities in Chicago (such as railroads) built a variety of movable bridges, including bascules, throughout the city. By now, the structures have become defining features of the urban landscape.

4-056. Aerial view of bascule bridges on the Chicago River looking east toward Lake Michigan. Jet T. Lowe, photographer, 1999. P&P,HAER,ILL,16-CHIG,137-12.

4-056

4-057. Cortland Street Bridge, Chicago River, Chicago, Illinois, 1902. John Ericson, city engineer, designer; American Bridge Company (Lassig Works, Chicago, IL), superstructure; Fitzsimons & Connell Company (Chicago, IL), substructure. Jet T. Lowe, photographer, 1987. P&P,HAER,ILL,16-CHIG,136-3.

This was the first of the simple trunnion bascule bridges, commonly known as Chicago-type bridges, developed by the Chicago Department of Public Works. The weight of each leaf balances on a trunnion (horizontal steel pivot) rotated by a rack and pinion gear driven by an electric motor (4-089, 4-090). The lifting machinery and counterweights are housed beneath the roadway. Ericson and his colleagues adapted the design from that of Tower Bridge in London, completed in 1894. By 1960, the department had over fifty such bridges in operation.

4-058. View of lift machinery showing 38-horsepower motor (far right) and rack and pinion gear (left), Cortland Street Bridge, Chicago River, Chicago, Illinois, 1902. John Ericson, city engineer, designer; American Bridge Company (Lassig Works, Chicago, IL), superstructure; Fitzsimons & Connell Company (Chicago, IL), substructure. Jet T. Lowe, photographer, 1987. P&P,HAER,ILL,16-CHIG,136-6.

4-059. View of lift mechanism showing hand-activated brake controlled by cable from operator's house, Cortland Street Bridge, Chicago River, Chicago, Illinois, 1902. John Ericson, city engineer, designer; American Bridge Company (Lassig Works, Chicago, IL), superstructure; Fitzsimons & Connell Company (Chicago, IL), substructure. Jet T. Lowe, photographer, 1987. P&P,HAER,ILL,16-CHIG,136-8.

4-057

4-058

4-059

4-060

4-061

4-060. Michigan Avenue Bridge, Chicago River, Chicago, Illinois, 1918–1920. Alexander von Babo, Chicago Bureau of Engineering, engineer; Edward Bennett, architect. Jet T. Lowe, photographer, 1987. P&P,HAER,ILL,16-CHIG,129-1.

The bridge has parallel double-leaf, double-deck spans that can be operated independently. The four monumental towers attest to its prominent role in shaping the image of Michigan Avenue as a grand urban boulevard envisioned in the master plan of 1909 proposed by Edward Bennett and Daniel Burnham. Bennett embellished the towers with bas-reliefs celebrating the city's history.

4-061. Michigan Avenue Bridge, Chicago River, Chicago, Illinois, 1918–1920. Alexander von Babo, Chicago Bureau of Engineering, engineer; Edward Bennett, architect. Jet T. Lowe, photographer, 1987. P&P,HAER,ILL,16-CHIG,129-3.

The bridge has a clear span between piers of 220 feet. The lower deck was intended for heavy truck traffic.

4-062. Tower, Michigan Avenue Bridge, Chicago River, Chicago, Illinois, 1918–1920. Alexander von Babo, Chicago Bureau of Engineering, engineer; Edward Bennett, architect. Jet T. Lowe, photographer, 1987. P&P,HAER,ILL,16-CHIG,129-9.

4-063. View of trunnion girder (riveted members, left) and counterweight (right, beyond railing), Michigan Avenue Bridge, Chicago River, Chicago, Illinois, 1918–1920. Alexander von Babo, Chicago Bureau of Engineering, engineer; Edward Bennett, architect. Jet T. Lowe, photographer, 1987. P&P,HAER,ILL,16-CHIG,129-12.

4-062

4-063

4-064

4-067

4-065

4-066

4-064. Wabash Avenue Bridge, Chicago River, Chicago, Illinois, 1930. Thomas G. Pihlfeldt, city bridge engineer, designer; Kelter & Elliot Company, builder. Jet T. Lowe, photographer, 1987. P&P,HAER,ILL,16-CHIG,133-3.

4-065. Control tower, Wabash Avenue Bridge, Chicago River, Chicago, Illinois, 1930. Thomas G. Pihlfeldt, city bridge engineer, designer; Kelter & Elliot Company, builder. Jet T. Lowe, photographer, 1987. P&P,HAER,ILL,16-CHIG,133-5.

4-066. Control room, Wabash Avenue Bridge, Chicago River, Chicago, Illinois, 1930. Thomas G. Pihlfeldt, city bridge engineer, designer; Kelter & Elliot Company, builder. Jet T. Lowe, photographer, 1987. P&P,HAER,ILL,16-CHIG,133-13.

4-067. Counterweight pit, Wabash Avenue Bridge, Chicago River, Chicago, Illinois, 1930. Thomas G. Pihlfeldt, city bridge engineer, designer; Kelter & Elliot Company, builder. Jet T. Lowe, photographer, 1987. P&P,HAER,ILL,16-CHIG,133-6.

4-068. Chicago & Alton Railroad Bridge, Chicago River, Chicago, Illinois, 1906. William M. Hughes, engineer; American Bridge Company, superstructure; Kelly & Atkinson Company, substructure. Martin Stupich, photographer, 1988. P&P,HAER,ILL,16-CHIG,114-1.

Among the many variations of the trunnion-type bascule was John W. Page's patented design that incorporated the counterweight into the structure of the approach span, which pivoted with the bascule leaf.

4-069. Eight-track bascule bridge, Pennsylvania Railroad, Chicago Sanitary & Ship Canal, Chicago, Illinois, 1901 (reconfigured 1909–1910). Scherzer Rolling Lift Bridge Company, 1901; Chicago Bridge & Iron Works, 1909–1910. Martin Stupich, photographer, 1988. P&P,HAER,ILL,16-CHIG,151-1.

4-070. Eastern spans showing counterweights and curved pivot girders, eight-track bascule bridge, Pennsylvania Railroad, Chicago Sanitary & Ship Canal, Chicago, Illinois, 1901 (reconfigured 1909–1910). Scherzer Rolling Lift Bridge Company, 1901; Chicago Bridge & Iron Works, 1909–1910. Martin Stupich, photographer, 1988. P&P,HAER,ILL,16-CHIG,151-4.

4-068

4-069

4-070

4-071

Bascule Bridges across the United States

4-071. Aerial view, Fort Point Channel rolling lift bridge, New York, New Haven & Hartford Railroad, Boston, Massachusetts, 1898–1900 (demolished 2003). William Scherzer, Scherzer Rolling Lift Bridge Company (Chicago, IL), engineer and builder. Jack E. Boucher, photographer, 1977. P&P,HAER,MASS,13-BOST,82-2.

William Scherzer (1858–1893) received a patent in 1893 for his design for bascule bridges that rolled the leaf on a rocking mechanism up and away from the river channel. His company became one of the leading builders of bascule spans in the early twentieth century. Scherzer's mechanism reduced the bearing points and encountered less friction than the trunnion of the Chicago-type bascule. It located the counterweight above the deck, allowing for a wider channel opening. The shifting center of gravity, however, added stress to the abutments, and the mechanism was more complicated and moved more slowly than the simple trunnion of the Chicago type. The Fort Point bridge was the first to be built by Scherzer's company outside Chicago. It consisted of six pairs of rails grouped on three independently operable spans crossing the channel at a 42-degree skew.

4-072. Fort Point Channel rolling lift bridge, New York, New Haven & Hartford Railroad, Boston, Massachusetts, 1898–1900 (demolished 2003). William Scherzer, Scherzer Rolling Lift Bridge Company (Chicago, IL), engineer and builder. Jet T. Lowe, photographer, 1982. P&P,HAER,MASS,13-BOST,82-3.

4-073. View of leaves in open position, Fort Point Channel rolling lift bridge, New York, New Haven & Hartford Railroad, Boston, Massachusetts, 1898–1900 (demolished 2003). William Scherzer, Scherzer Rolling Lift Bridge Company (Chicago, IL), engineer and builder. DETR, ca. 1904. P&P,LC-D4-17090.

4-072

4-073

4-074. View showing counterweights and curved roller arm, Fort Point Channel rolling lift bridge, New York, New Haven & Hartford Railroad, Boston, Massachusetts, 1898–1900 (demolished 2003). William Scherzer, Scherzer Rolling Lift Bridge Company (Chicago, IL), engineer and builder. Jet T. Lowe, photographer, 1982. P&P,HAER,MASS,13-BOST,82-6.

4-075. Seddon Island Bridge, Garrison Channel, Tampa, Florida, 1908–1909 (demolished 1982). Designed using Scherzer rolling lift patents and built by Phoenix Bridge Company (Phoenixville, PA). Unidentified photographer, 1981. P&P,HAER,FLA,29-TAMP, 21-2.

4-076. Lifting gears, Seddon Island Bridge, Garrison Channel, Tampa, Florida, 1908–1909 (demolished 1982). Designed using Scherzer rolling lift patents and built by Phoenix Bridge Company (Phoenixville, PA). Unidentified photographer, 1981. P&P,HAER, FLA,29-TAMP,21-23.

4-074

4-075

4-076

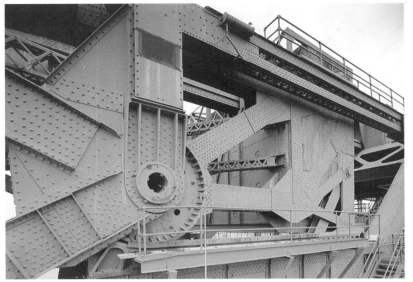

4-077

4-078

4-079

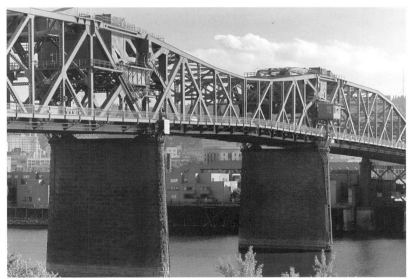

4-080

4-077. Hinge mechanism in closed position, Broadway Bridge, Willamette River, Portland, Oregon, 1912–1913. Ralph Modjeski, chief engineer; Strobel Steel Construction Company (Chicago, IL), patent holder for Rall lift mechanism; Pennsylvania Steel Company (Steelton, PA), superstructure; Union Bridge Company (Kansas City, MO), substructure. Jet T. Lowe, photographer, 1990. P&P,HAER,ORE,26-PORT,9-7.

4-078. Hinge mechanism in open position, Broadway Bridge, Willamette River, Portland, Oregon, 1912–1913. Ralph Modjeski, chief engineer; Strobel Steel Construction Company (Chicago, IL), patent holder for Rall lift mechanism; Pennsylvania Steel Company (Steelton, PA), superstructure; Union Bridge Company (Kansas City, MO), substructure. Jet T. Lowe, photographer, 1990. P&P,HAER,ORE,26-PORT,9-9.

4-079. Broadway Bridge in open position, Willamette River, Portland, Oregon, 1912–1913. Ralph Modjeski, chief engineer; Strobel Steel Construction Company (Chicago, IL), patent holder for Rall lift mechanism; Pennsylvania Steel Company (Steelton, PA), superstructure; Union Bridge Company (Kansas City, MO), substructure. Jet T. Lowe, photographer, 1990. P&P,HAER,ORE,26-PORT,9-15.

4-080. Broadway Bridge, Willamette River, Portland, Oregon, 1912–1913. Ralph Modjeski, chief engineer; Strobel Steel Construction Company (Chicago, IL), patent holder for Rall lift mechanism; Pennsylvania Steel Company (Steelton, PA), superstructure; Union Bridge Company (Kansas City, MO), substructure. Jet T. Lowe, photographer, 1990. P&P,HAER,ORE,26-PORT,9-12.

This is one of six bascule bridges in the United States built with the roller lift mechanism patented by Theodore Rall in 1901. Like Scherzer-type bridges, it raises the leaf up and away from the river channel, but it rotates on a wheel and track rather than on a curved girder.

4-081

4-082

4-081. Darby River Bridge, Philadelphia, Baltimore & Washington Railroad, Essington, Delaware, Pennsylvania, 1917–1918. Strauss Bascule Bridge Company (Chicago, IL), engineer; Bethlehem Steel Bridge Corporation (Steelton, PA), builder. Jet T. Lowe, photographer, 1999. P&P,HAER,PA,23-EDDY,1-7.

Early in his career, Joseph Boorman Strauss (1870–1938), best known as the chief engineer of the Golden Gate Bridge (see 5-095–5-101), founded a company in 1902 that specialized in the design of bascule bridges.

4-082. Detail showing trunnion and counterweight in open position, Darby River Bridge, Philadelphia, Baltimore & Washington Railroad, Essington, Delaware, Pennsylvania, 1917–1918. Strauss Bascule Bridge Company (Chicago, IL), engineer; Bethlehem Steel Bridge Corporation (Steelton, PA), builder. Jet T. Lowe, photographer, 1999. P&P,HAER,PA,23-EDDY,1-6.

4-083. Saugus River Drawbridge, Boston & Maine Railroad, Saugus, Essex County, Massachusetts, 1911. Strauss Bascule Company (Chicago, IL), engineer; Phoenix Bridge Company (Phoenixville, PA), builder. Abbot Boyle, Inc., photographer, 1987. P&P,HAER,MASS,5-SAUG,2-5.

4-084. Military Street Bridge, Black River, Port Huron, St. Clair County, Michigan, 1912–1913. Strauss Bascule Bridge Company (Chicago, IL), Osborne Engineering Company (Cleveland, OH), engineers; McKenzie Bridge Company (Port Huron, MI), Detroit Bridge & Steel Works (River Rouge, MI), builders. Carla Anderson, photographer, 1990. P&P,HAER,MICH,74-POHU,2-11.

4-083

4-084

4-085

4-086

4-087

4-085. Mystic River Bridge, Groton, New London County, Connecticut, 1922. Waddell & Son, Thomas E. Brown & Son, engineers; American Bridge Company (Pittsburgh, PA), fabricator; J. E. Fitzgerald Company (New London, CT), builder. Wayne Fleming, photographer, 1996. P&P,HAER,CONN,6-GROT,2-1.

Thomas E. Brown developed an overhead counterweight system known as the Brown Balance Beam that reduced the movement of the weights and offered advantages in the way the leaf moved and locked.

4-086. Counterweights and operating wheel, Mystic River Bridge, Groton, New London County, Connecticut, 1922. Waddell & Son, Thomas E. Brown & Son, engineers; American Bridge Company (Pittsburgh, PA), fabricator; J. E. Fitzgerald Company (New London, CT), builder. Wayne Fleming, photographer, 1996. P&P,HAER, CONN,6-GROT,2-15.

4-087. Trunnion, Mystic River Bridge, Groton, New London County, Connecticut, 1922. Waddell & Son, Thomas E. Brown & Son, engineers; American Bridge Company (Pittsburgh, PA), fabricator; J. E. Fitzgerald Company (New London, CT), builder. Wayne Fleming, photographer, 1996. P&P,HAER,CONN,6-GROT,2-14.

4-088. Montlake Bridge, Lake Washington Ship Canal, Seattle, Washington, 1925. Seattle City Engineering Department, designer; Edgar Blair, Harlan Thomas, A. W. Albertson, consulting architects; Wallace Equipment Company (Seattle, WA), superstructure; C. L. Creelman Company (Seattle, WA), substructure. Jet T. Lowe, photographer, 1993. P&P,HAER,WASH,17-SEAT,14-1.

4-089. Axonometric view of lift mechanism, Montlake Bridge, Lake Washington Ship Canal, Seattle, Washington, 1925. Seattle City Engineering Department, designer; Edgar Blair, Harlan Thomas, A. W. Albertson, consulting architects; Wallace Equipment Company (Seattle, WA), superstructure; C. L. Creelman Company (Seattle, WA), substructure. Catherine I. Kudlik, delineator, 1993. P&P,HAER,WASH,17-SEAT,14-,sheet no. 2.

The Seattle engineers based their design on the Chicago-type bascule.

4-090. Section of bascule leaf, axonometric views of lock mechanism, Montlake Bridge, Lake Washington Ship Canal, Seattle, Washington, 1925. Seattle City Engineering Department, designer; Edgar Blair, Harlan Thomas, A. W. Albertson, consulting architects; Wallace Equipment Company (Seattle, WA), superstructure; C. L. Creelman Company (Seattle, WA), substructure. Catherine I. Kudlik, delineator, 1993. P&P,HAER,WASH,17-SEAT,14-,sheet no. 3.

4-088

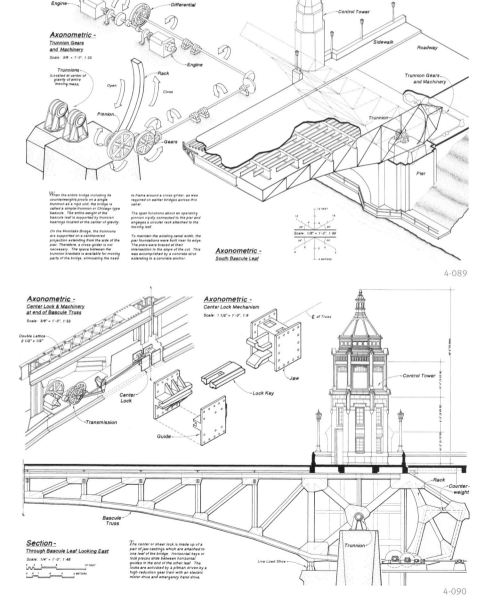

4-089

4-090

4-091

4-091. View of counterweight (left), bull gear, and trunnion, Montlake Bridge, Lake Washington Ship Canal, Seattle, Washington, 1925. Seattle City Engineering Department, designer; Edgar Blair, Harlan Thomas, A. W. Albertson, consulting architects; Wallace Equipment Company (Seattle, WA), superstructure; C. L. Creelman Company (Seattle, WA), substructure. Jet T. Lowe, photographer, 1993. P&P,HAER,WASH,17-SEAT,14-13.

4-092. Final reduction gears, Montlake Bridge, Lake Washington Ship Canal, Seattle, Washington, 1925. Seattle City Engineering Department, designer; Edgar Blair, Harlan Thomas, A. W. Albertson, consulting architects; Wallace Equipment Company (Seattle, WA), superstructure; C. L. Creelman Company (Seattle, WA), substructure. Jet T. Lowe, photographer, 1993. P&P,HAER,WASH,17-SEAT,14-18.

4-092

4-093. University Avenue Bridge, Schuylkill River, Philadelphia, Pennsylvania, 1927–1933. John A. Vogelson, Steven H. Noyes, Philadelphia Department of Public Works, engineers; Paul Cret, consulting architect; Dravo Contracting Company (Pittsburgh, PA), builder. Jet T. Lowe, photographer, 1999. P&P,HAER,PA,51-PHILA,730-3.

4-094. Underside of floor system, University Avenue Bridge, Schuylkill River, Philadelphia, Pennsylvania, 1927–1933. John A. Vogelson, Steven H. Noyes, Philadelphia Department of Public Works, engineers; Paul Cret, consulting architect; Dravo Contracting Company (Pittsburgh, PA), builder. Jet T. Lowe, photographer, 1999. P&P,HAER,PA,51-PHILA,730-9.

4-095. Chicago-style trunnion, University Avenue Bridge, Schuylkill River, Philadelphia, Pennsylvania, 1927–1933. John A. Vogelson, Steven H. Noyes, Philadelphia Department of Public Works, engineers; Paul Cret, consulting architect; Dravo Contracting Company (Pittsburgh, PA), builder. Jet T. Lowe, photographer, 1999. P&P,HAER,PA,51-PHILA,730-10.

4-093

4-094

4-095

Lift bridges require towers and more complex lifting machinery than bascule bridges, but their spans can be significantly longer, unhampered by the limitations imposed by the latter's cantilevered arms, bearing capacity of the trunnions, and arrangement of counterweights.

4-096

4-097

4-096. South Halsted Street Lift Bridge, Chicago River, Chicago, Illinois, 1894 (demolished). J. A. L. Waddell, engineer. DETR, ca. 1900–1915. P&P,LC-D4-500193.

This was the first long-span, high-rise vertical lift bridge built in the United States. Waddell, the foremost designer of such structures at the turn of the twentieth century, based it on an unrealized design he had made in 1892 for Duluth, Minnesota.

4-097. Oswego Canal Railroad Lift Bridge, New York, West Shore & Buffalo Railroad, Syracuse, Onondaga County, 1883. Albert Lucius, engineer; Hilton Bridge Company (Albany, NY), builder. Cover illustration, Scientific American 49, no. 20 (October 1883). P&P,LC-USZ62-61617.

4-098. Aerial Lift Bridge, Duluth Ship Canal, Duluth, St. Louis County, Minnesota, 1905 (as aerial transfer bridge), transformed 1929–1930 (as vertical lift bridge). Thomas F. McGilvray, C. A. P. Turner, engineers. DETR, photographer, ca. 1905. P&P,LC-D4-18810.

Built more than a decade after J. A. L. Waddell's project for the crossing, the bridge conveyed pedestrians, horse-drawn wagons, and automobiles across the 300-foot channel on a large platform suspended from the high truss. The aerial transfer bridge type, also known as a transporter bridge, was invented in the early 1890s by the French engineer Ferdinand Arnodin (1845–1924). Approximately a dozen were built in Europe at the turn of the twentieth century (the oldest in Portugalete, Spain, 1893, remains in use), but only one other was built in the United States, the Sky Ride at the Century of Progress Exposition, Chicago, 1933 (demolished).

4-099. Roadway in lower position, Aerial Lift Bridge, Duluth Ship Canal, Duluth, St. Louis County, Minnesota, 1905 (as aerial transfer bridge), transformed 1929–1930 (as vertical lift bridge). Thomas F. McGilvray, C. A. P. Turner, engineers. Jet T. Lowe, photographer, P&P,HAER,MINN,69-DULU,9-5.

McGilvray and Turner transformed the structure into a vertical lift bridge by strengthening the towers and upper truss and replacing the suspended platform with a truss and hoist mechanism.

4-100. Roadway in upper position, Aerial Lift Bridge, Duluth Ship Canal, Duluth, St. Louis County, Minnesota, 1905 (as aerial transfer bridge), transformed 1929–1930 (as vertical lift bridge). Thomas F. McGilvray, C. A. P. Turner, engineers. Jet T. Lowe, photographer, P&P,HAER,MINN,69-DULU,9-3.

4-098

4-099

4-100

4-101

4-102

4-103

4-101. Hawthorne Bridge, Willamette River, Portland, Oregon, 1909–1910 (rehabilitated 1999). Waddell & Harrington (Kansas City, MO), engineers; Pennsylvania Steel Company (Steelton, PA), fabricator; United Engineering & Construction Company (Portland, OR), superstructure; Robert Wakefield & Company (Portland, OR), substructure. Jet T. Lowe, photographer, 1990. P&P,HAER,ORE,26-PORT,10-6.

This was Waddell & Harrington's third and remains their oldest extant vertical lift bridge. Waddell's partner, John Lyle Harrington (1868–1942), contributed to the design the former had introduced with the South Halsted Street Bridge (4-096) by improving the hoist mechanism. The Hawthorne Bridge's 250-foot-long lift span can be raised 110 feet, creating a vertical clearance of 160 feet at mean low water.

4-102. Lift cable connections at tower, Hawthorne Bridge, Willamette River, Portland, Oregon, 1909–1910 (rehabilitated 1999). Waddell & Harrington (Kansas City, MO), engineer; Pennsylvania Steel Company (Steelton, PA), fabricator; United Engineering & Construction Company (Portland, OR), superstructure; Robert Wakefield & Company (Portland, OR), substructure. Jet T. Lowe, photographer, 1990. P&P,HAER,ORE,26-PORT,10-8.

4-103. Engine room and reduction gears atop central span, Hawthorne Bridge, Willamette River, Portland, Oregon, 1909–1910 (rehabilitated 1999). Waddell & Harrington (Kansas City, MO), engineers; Pennsylvania Steel Company (Steelton, PA), fabricator; United Engineering & Construction Company (Portland, OR), superstructure; Robert Wakefield & Company (Portland, OR), substructure. Jet T. Lowe, photographer, 1990. P&P,HAER,ORE,26-PORT, 10-11.

4-104. Aerial view showing lift span in raised position, Armour, Swift, Burlington Bridge, Missouri River, Kansas City, Missouri, 1910–1912 (roadway removed ca. 1982). Waddell & Harrington (Kansas City, MO), engineers; McClintic-Marshall Construction Company (Pittsburgh, PA), superstructure; James O'Connor & Son, substructure. James K. Corrigan, Bud Cox, photographers, 1981. P&P,HAER,MO,48-KANCI,16-7.

The lower railroad deck lifts into the through truss and nests beneath the roadway.

4-105. Railroad deck showing hangers and underside of roadway deck, Armour, Swift, Burlington Bridge, Missouri River, Kansas City, Missouri, 1910–1912 (roadway removed ca. 1982). Waddell & Harrington (Kansas City, MO), engineers; McClintic-Marshall Construction Company, (Pittsburgh, PA) superstructure; James O'Connor & Son, substructure. James K. Corrigan, Bud Cox, photographers, 1981. P&P,HAER,MO,48-KANCI,16-12.

4-106. Counterweights and bridge tender's house on rail deck, Armour, Swift, Burlington Bridge, Missouri River, Kansas City, Missouri, 1910–1912 (roadway removed ca. 1982). Waddell & Harrington (Kansas City, MO), engineers; McClintic-Marshall Construction Company (Pittsburgh, PA), superstructure; James O'Connor & Son, substructure. James K. Corrigan, Bud Cox, photographers, 1981. P&P,HAER,MO,48-KANCI,16-29.

4-107

4-108

4-109

4-107. Steel Bridge, Oregon Railroad & Navigation Co., Willamette River, Portland, Oregon, 1910–1912. Waddell & Harrington (Kansas City, MO), designers; American Bridge Company (Pittsburgh, PA), fabricator; Union Bridge & Construction Company (Kansas City, MO), superstructure; Robert Wakefield & Company (Portland, OR), substructure. Orin Russey, photographer, 1998. P&P,HAER,ORE,26-PORT,14-11.

Waddell & Harrington designed this bridge so that (1) the lower rail and upper roadway decks could be raised together for ships requiring high clearances, or (2) the rail deck could be raised independently to accommodate smaller vessels without disrupting highway traffic.

4-108. Snowden Bridge, Great Northern Railroad, Missouri River, Nohly Vic., Richland County, Montana, 1913. Waddell & Harrington (Kansas City, MO), engineers; American Bridge Company (Gary, IN), fabricator; Union Bridge & Construction Company (Kansas City, MO), builder. Jet T. Lowe, photographer, 1980. P&P,HAER,MONT,42-NOH.V,1-1.

4-109. Interior of hoist house, Snowden Bridge, Great Northern Railroad, Missouri River, Nohly Vic., Richland County, Montana, 1913. Waddell & Harrington (Kansas City, MO), engineers; American Bridge Company (Gary, IN), fabricator; Union Bridge & Construction Company (Kansas City, MO), builder. Jet T. Lowe, photographer, 1980. P&P,HAER,MONT,42-NOH.V,1-18.

4-110. South tower, Snowden Bridge, Great Northern Railroad, Missouri River, Nohly Vic., Richland County, Montana, 1913. Waddell & Harrington (Kansas City, MO), engineers; American Bridge Company (Gary, IN), fabricator; Union Bridge & Construction Company (Kansas City, MO), builder. Jet T. Lowe, photographer, 1980. P&P,HAER,MONT,42-NOH.V,1-13.

4-110

4-111. Hackensack River vertical lift bridges, Kearny, Hudson County, New Jersey. Jack E. Boucher, photographer, 1978. P&P,HAER,NJ,9-KEAR,3-3.

From left to right are the the Erie & Lackawanna Railroad Bridge, the Newark Turnpike Bridge, the Conrail Bridge, and the PATH Transit System Bridge. At top right is a portion of the Pulaski Skyway (see 3-368).

4-112. Newark Bay Bridge, Central Railroad of New Jersey, Newark, Essex County, New Jersey, 1924–1925 (demolished 1980). J. A. L. Waddell, engineer. U.S. Coast Guard, photographer, 1979. P&P,HAER,NJ,7-NEARK, 16-40.

4-113. Newark Bay Bridge showing passage of the Sealand Galloway container ship, Central Railroad of New Jersey, Newark, Essex County, New Jersey, 1924–1925 (demolished 1980). J. A. L. Waddell, engineer. U.S. Coast Guard, photographer, 1979. P&P,HAER,NJ,7-NEARK,16-48.

4-111

4-112

4-113

4-114

4-114. St. Francis River Bridge, Lake City, Craighead County, Arkansas, 1934. Arkansas State Highway Commission, designer; Vincennes Bridge Company (Vincennes, IN), builder. Louise T. Cawood, photographer, 1988. P&P,HAER,ARK,16-LACI,1-4.

4-115. Cape Cod Canal Railroad Bridge, Buzzards Bay, Barnstable County, Massachusetts, 1933–1935. Parsons, Klapp, Brinkerhoff & Douglas (New York, NY), engineers; McKim, Mead & White, consulting architects. Martin Stupich, photographer, 1987. P&P,HAER,MASS,1-BUZBA,1-2.

In the bridge's raised position, the 544-foot-long through truss and supporting towers become a monumental portal to the ship canal.

4-116. Lowered position, Cape Cod Canal Railroad Bridge, Buzzards Bay, Barnstable County, Massachusetts, 1933–1935. Parsons, Klapp, Brinkerhoff & Douglas (New York, NY), engineers; McKim, Mead & White, consulting architects. Martin Stupich, photographer, 1987. P&P,HAER,MASS,1-BUZBA,1-5.

4-115

4-116

4-117. Aerial view, Sacramento River Bridge (Tower Bridge), Sacramento, California, 1934–1935. California Division of Highways, engineer; Alfred Eichler, consulting architect. Don Tateishi, photographer, 1985. P&P,HAER,CAL, 34-SAC,58-5.

This bridge was designed as a monumental entrance to the state capital for traffic on transcontinental Route 40 and light rail. Steel plates painted silver (now ochre) clad the structure in the streamlined style then popular as a symbol of progress and modernity.

4-118. Pylon that originally anchored trolley wires, Sacramento River Bridge (Tower Bridge), Sacramento, California, 1934–1935. California Division of Highways, engineer; Alfred Eichler, consulting architect. Don Tateishi, photographer, 1985. P&P,HAER, CAL,34-SAC,58-12.

4-119. Sacramento River Bridge (Tower Bridge), Sacramento, California, 1934–1935. California Division of Highways, engineer; Alfred Eichler, consulting architect. Don Tateishi, photographer, 1985. P&P,HAER, CAL,34-SAC,58-10.

4-117

4-118

4-119

4-120

4-120. Portal looking toward Sacramento, Sacramento River Bridge (Tower Bridge), Sacramento, California, 1934–1935. California Division of Highways, engineer; Alfred Eichler, consulting architect. Don Tateishi, photographer, 1985. P&P,HAER,CAL,34-SAC,58-8.

4-121. Engine house, Sacramento River Bridge (Tower Bridge), Sacramento, California, 1934–1935. California Division of Highways, engineer; Alfred Eichler, consulting architect. Don Tateishi, photographer, 1985. P&P,HAER,CAL,34-SAC,58-16.

4-122. Interior of engine house showing electric motors and transmission, Sacramento River Bridge (Tower Bridge), Sacramento, California, 1934–1935. California Division of Highways, engineer; Alfred Eichler, consulting architect. Don Tateishi, photographer, 1985. P&P,HAER,CAL,34-SAC,58-21.

4-121

4-122

4-123

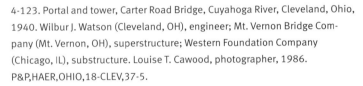

4-124

4-123. Portal and tower, Carter Road Bridge, Cuyahoga River, Cleveland, Ohio, 1940. Wilbur J. Watson (Cleveland, OH), engineer; Mt. Vernon Bridge Company (Mt. Vernon, OH), superstructure; Western Foundation Company (Chicago, IL), substructure. Louise T. Cawood, photographer, 1986. P&P,HAER,OHIO,18-CLEV,37-5.

4-124. View of movable span and engine house from tower, Carter Road Bridge, Cuyahoga River, Cleveland, Ohio, 1940. Wilbur J. Watson (Cleveland, OH), engineer; Mt. Vernon Bridge Company (Mt. Vernon, OH), superstructure; Western Foundation Company (Chicago, IL), substructure. Louise T. Cawood, photographer, 1986. P&P,HAER,OHIO,18-CLEV,37-11.

4-125. Bearing on abutment, Carter Road Bridge, Cuyahoga River, Cleveland, Ohio, 1940. Wilbur J. Watson (Cleveland, OH), engineer; Mt. Vernon Bridge Company (Mt. Vernon, OH), superstructure; Western Foundation Company (Chicago, IL), substructure. Louise T. Cawood, photographer, 1986. P&P,HAER,OHIO,18-CLEV,37-8.

4-125

SUSPENSION AND CABLE-STAYED BRIDGES

SUSPENSION BRIDGES

Suspension bridges have decks supported from cables anchored at each end of the span. Typically, the cables pass over towers to allow optimal arrangements for the height of the deck with respect to the approaches and clearance for traffic passing below.

Ancient civilizations in China, India, and pre-Columbian South America built suspension bridges with cables woven from grasses and other fibrous plants, but medieval and early modern Europeans showed little interest in the type despite their familiarity with the analogous technology of rigging sailing vessels. The decisive factor was cultural. For Europeans, bridges were solid, and builders sought structural solutions, such as the arch, that offered maximum rigidity.

The history of modern suspension bridges began in the late eighteenth century in Britain and the United States and spread to the European continent, where in the early nineteenth century engineers such as Claude-Louis-Marie-Henri Navier (1785–1836) published theories quantifying their behavior. At the same time, advances in iron technology made it possible to economically fabricate cables of wire or chain capable of supporting long-span structures that could be built without the expensive and time-consuming centering required by most types of trusses and arches. These theoretical and technological innovations were observed firsthand in Europe by Charles Ellet Jr. (5-013–5-020) and John Augustus Roebling (5-021–5-047), who emerged as the leading designers of suspension bridges in America in the 1840s.

The size, elegance, and often dramatic settings of suspension bridges have sparked the imaginations of engineers and the public alike, and some of the designers of the largest of these structures enjoyed a degree of public attention beyond that accorded other prominent engineers.

5-001

Suspension Bridge Components

Towers, primary cables, suspenders supporting the deck, and anchorages are the principal components of suspension bridges.

5-001. Boldman Bridge, Levisa Fork, Pikeville Vic., Pike County, Kentucky, ca. 1935. G. D. Rawlings and J. E. Daniel II, photographers, 1984. P&P,HAER,KY,98-PIKVI.V,1-1.

At 419 feet long and 9 feet wide, the Boldman Bridge is representative of a type of suspension bridge, once common in the hilly regions of Kentucky, known as a swinging bridge because of its lively action due to lightweight construction and the absence of stiffening trusses reinforcing the deck.

5-002

5-003

5-002. Main cable, suspender, Boldman Bridge, Levisa Fork, Pikeville Vic., Pike County, Kentucky, ca. 1935. G. D. Rawlings and J. E. Daniel II, photographers, 1984. P&P,HAER,KY,98-PIKVI.V,1-8.

5-003. Deck and suspender, Boldman Bridge, Levisa Fork, Pikeville Vic., Pike County, Kentucky, ca. 1935. G. D. Rawlings and J. E. Daniel II, photographers, 1984. P&P,HAER,KY,98-PIKVI.V,1-7.

5-004. Anchorage, Boldman Bridge, Levisa Fork, Pikeville Vic., Pike County, Kentucky, ca. 1935. G. D. Rawlings and J. E. Daniel II, photographers, 1984. P&P,HAER,KY,98-PIKVI.V,1-11.

5-005. Deck, Swinging Bridge, Hazard, Perry County, Kentucky. Marion Post Wolcott, photographer, 1944. P&P,LC-USF34-055733-D.

Unlike the Boldman Bridge, the Swinging Bridge has a deck stiffened by a wooden truss.

5-004

5-005

5-006

5-007

5-008

5-006. George Washington Bridge, Hudson River, New York, New York, 1927–1931; lower deck, 1958–1962. Othmar H. Ammann, chief engineer; Allston Dana, design engineer; Cass Gilbert, architect; McClintic-Marshall Company (Pittsburgh, PA), superstructure; John A. Roebling's Sons Company (Trenton, NJ), cables; Bethlehem Steel Company (Bethlehem, PA), lower deck. Jack E. Boucher, photographer, 1978. P&P,HAER,NY,31-NEYO,161-13.

One of the outstanding engineering works of the twentieth century, the George Washington Bridge has a suspended span 3,500 feet long and 120 feet wide. At the time of construction, its clear span was more than twice that of any other bridge. Its principal engineers, Othmar Ammann (1879–1965) and Allston Dana (1884—1952), were among the leading designers of suspension bridges in the mid-twentieth century (5-103–5-110).

5-007. View of main cables, suspenders, and tower, George Washington Bridge, Hudson River, New York, New York, 1927–1931. Othmar H. Ammann, chief engineer; Allston Dana, design engineer; Cass Gilbert, architect. Margaret Bourke-White, photographer, ca. 1933. P&P,LC-USZ62-76063.

5-008. Tower and double deck, George Washington Bridge, Hudson River, New York, New York, 1927–1931; lower deck, 1958—1962. Othmar H. Ammann, chief engineer; Allston Dana, design engineer; Cass Gilbert, architect; McClintic-Marshall Company (Pittsburgh, PA), superstructure; John A. Roebling's Sons Company (Trenton, NJ), cables; Bethlehem Steel Company (Bethlehem, PA), lower deck. Jet T. Lowe, photographer, 1986. P&P,HAER,NY,31-NEYO, 161-18.

5-009

5-009. Main cable, suspender, George Washington Bridge, Hudson River, New York, New York, 1927–1931; lower deck, 1958–1962. Othmar H. Ammann, chief engineer; Allston Dana, design engineer; Cass Gilbert, architect; McClintic-Marshall Company (Pittsburgh, PA), superstructure; John A. Roebling's Sons Company (Trenton, NJ), cables; Bethlehem Steel Company (Bethlehem, PA), lower deck. Jet T. Lowe, photographer, 1986. P&P,HAER,NY,31-NEYO,161-29.

5-010. Tower arch and stiffening trusses underneath lower deck, George Washington Bridge, Hudson River, New York, New York, 1927–1931; lower deck, 1958–1962. Othmar H. Ammann, chief engineer; Allston Dana, design engineer; Cass Gilbert, architect; McClintic-Marshall Company (Pittsburgh, PA), superstructure; John A. Roebling's Sons Company (Trenton, NJ), cables; Bethlehem Steel Company (Bethlehem, PA), lower deck. Jet T. Lowe, photographer, 1986. P&P,HAER,NY,31-NEYO,161-34.

5-011. Anchorage, George Washington Bridge, Hudson River, New York, New York, 1927–31; lower deck, 1958–1962. Othmar H. Ammann, chief engineer; Allston Dana, design engineer; Cass Gilbert, architect; McClintic-Marshall Company (Pittsburgh, PA), superstructure; John A. Roebling's Sons Company (Trenton, NJ), cables; Bethlehem Steel Company (Bethlehem, PA), lower deck. Jet T. Lowe, photographer, 1986. P&P,HAER,NY,31-NEYO,161-54.

5-010

5-011

Nineteenth-Century Suspension Bridges

5-012. Chain Bridge at Little Falls, Potomac River, McLean, Fairfax County, Virginia, 1810 (demolished). James Finley, designer. Augustus Kollner, wash drawing, 1839. P&P,LC-USZ61-1344.

James Finley (1756–1828), a judge in Fayette County in southwestern Pennsylvania, developed a system of iron chains and suspenders supporting a level roadbed that he patented in 1808. Between 1801 and 1816, he designed approximately 40 bridges in Pennsylvania, Virginia, and Maryland, none of which survive today. Some components of the last known bridge built under his patent in 1826 have been preserved at the Lehigh Gap in Palmerton, Pennsylvania. Finley's pioneering work contributed to the subsequent development of suspension span technology in Europe.

SUSPENSION BRIDGES BY CHARLES ELLET JR.

5-013. Fairmount Park suspension bridge, Schuylkill River, Philadelphia, Pennsylvania, 1841–1842 (demolished). Charles Ellet Jr., engineer. W. Croome, delineator, Graham's Magazine 20 (June 1842). P&P,4384 G.

Charles Ellet Jr. (1810–1862) studied civil engineering in France and was familiar with the analysis of suspension bridge behavior introduced by Claude-Louis-Marie-Henri Navier and the techniques for using wire cables instead of iron chains developed by Marc Seguin (1786–1875) and Louis Vicat (1786–1861). His 357-foot Schuylkill River crossing was the first successful wire-cable suspension bridge in the United States.

5-014. View of Wheeling Suspension Bridge from the south, Ohio River, Wheeling, Ohio County, West Virginia, 1847–1849. Charles Ellet Jr., engineer. Engraving in Charles A. Dana, The United States Illustrated (New York: H. J. Meyer, 1855). P&P,LC-USZ62-20541.

Nearly three times the length of his Fairmount Bridge in Philadelphia completed five years earlier, the span of Ellet's 1,010-foot Wheeling Suspension Bridge was by far the longest in the world. Its achievement was vital to the aspirations of Wheeling's citizens, who tied the city's prosperity to its role as a transportation hub linking the Ohio River, the National Road, and the B&O Railroad.

5-012

5-013

5-014

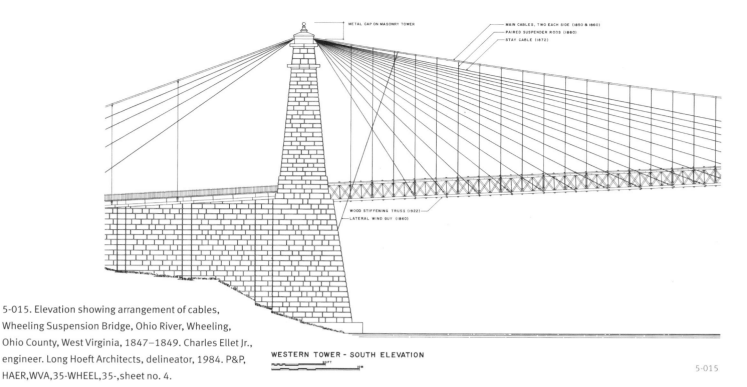

5-015. Elevation showing arrangement of cables, Wheeling Suspension Bridge, Ohio River, Wheeling, Ohio County, West Virginia, 1847–1849. Charles Ellet Jr., engineer. Long Hoeft Architects, delineator, 1984. P&P, HAER,WVA,35-WHEEL,35-,sheet no. 4.

WESTERN TOWER - SOUTH ELEVATION

5-015

A storm destroyed the suspended span in 1854. Ellet and William K. McComas quickly reopened it to one lane of traffic and reconstructed it six years later, incorporating features adapted from John Roebling's Niagara River bridge (5-030). In 1871–1872, Washington Roebling stiffened the span with stay cables.

5-016. View of cables showing stay cables and hangers, Wheeling Suspension Bridge, Ohio River, Wheeling, Ohio County, West Virginia, 1847–1849. Charles Ellet Jr., engineer. Jack E. Boucher, photographer, 1977. P&P,HAER,WVA,35-WHEEL,35-52.

5-017. View from western shore, Wheeling Suspension Bridge, Ohio River, Wheeling, Ohio County, West Virginia, 1847–1849. Charles Ellet Jr., engineer. Jack E. Boucher, photographer, 1977. P&P,HAER,WVA,35-WHEEL,35-56.

5-017

5-016

5-018

5-019

5-020

5-018. Cable saddle at east tower, Wheeling Suspension Bridge, Ohio River, Wheeling, Ohio County, West Virginia, 1847–1849. Charles Ellet Jr., engineer. Jack E. Boucher, photographer, 1977. P&P,HAER,WVA,35-WHEEL,35-41.

5-019. Cable anchor at the north side of the east tower, Wheeling Suspension Bridge, Ohio River, Wheeling, Ohio County, West Virginia, 1847–1849. Charles Ellet Jr., engineer. Jack E. Boucher, photographer, 1977. P&P,HAER,WVA,35-WHEEL,35-35.

5-020. Deck stiffening truss, Wheeling Suspension Bridge, Ohio River, Wheeling, Ohio County, West Virginia, 1847–1849. Charles Ellet Jr., engineer. Jack E. Boucher, photographer, 1977. P&P,HAER,WVA,35-WHEEL,35-34.

This wooden Howe truss with wrought-iron tension rods and cast-iron fittings at the joints likely was built during the reconstruction around 1860. The steel deck was built in 1956. The bridge was rehabilitated in 1999.

5-021

SUSPENSION BRIDGES BY JOHN A. ROEBLING

5-021. Delaware Aqueduct, Delaware & Hudson Canal, Delaware River, Lackawaxen, Pike County, Pennsylvania, and Minisink Ford, Sullivan County, New York, 1847–1848. John Roebling, engineer. Jet T. Lowe, photographer, 1989. P&P,HAER,PA,52-LACK,1-39.

John Augustus Roebling (1806–1869) emigrated from Prussia in 1831 with a civil engineering degree, but he began his life in western Pennsylvania as a farmer. Six years later, he turned to surveying routes for the region's rapidly expanding canal system. His experience with the mechanics of portaging canal boats over the Allegheny Mountains with inclined planes and cables inspired him to become the nation's first manufacturer of wire rope, which he subsequently applied to the design of suspension bridges. Among his early works were four suspension aqueducts commissioned by the Delaware & Hudson Canal Company to convey canal boats across rivers. Of these, only the four-span, 535-foot Delaware Aqueduct remains. It is the oldest surviving suspension bridge in the United States. Restoration was completed in 1995.

5-022

5-022. Toll house, Delaware Aqueduct, Delaware & Hudson Canal, Delaware River, Minisink Ford, Sullivan County, New York, 1908. Charles Spruks, builder. Jet T. Lowe, photographer, 1989. P&P,HAER,NY,53-MINFO,1-1.

Abandoned in 1898, the aqueduct was purchased nine years later by Charles Spruks, who converted it to a private highway bridge and added the toll house.

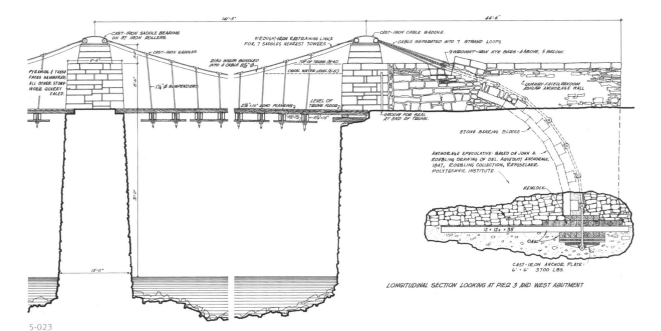

5-023

5-023. Longitudinal section, Delaware Aqueduct, Delaware & Hudson Canal, Delaware River, Lackawaxen, Pike County, Pennsylvania, and Minisink Ford, Sullivan County, New York, 1847–1848. John Roebling, engineer. Eric DeLony, Robert M. Vogel, delineators, 1969. P&P,HAER,PA,52-LACK,1-,sheet no. 3.

5-024. Isometric view reconstructing appearance of aqueduct in 1849, Delaware Aqueduct, Delaware & Hudson Canal, Delaware River, Lackawaxen, Pike County, Pennsylvania, and Minisink Ford, Sullivan County, New York, 1847–1848. John Roebling, engineer. Scott Barber, Brian D. Bartholomew, Elizabeth F. Knowlan, Anne Guerette, Dana Lockett, delineators, 1988–1891. P&P,HAER,PA,52-LACK,1-,sheet no. 3.

5-025. Cable saddle and anchorage, Delaware Aqueduct, Delaware & Hudson Canal, Delaware River, Lackawaxen, Pike County, Pennsylvania, and Minisink Ford, Sullivan County, New York, 1847–1848. John Roebling, engineer. Jet T. Lowe, photographer, 1989. P&P,HAER,PA,52-LACK,1-13.

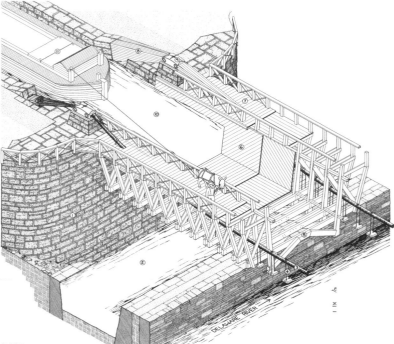

5-024

5-025

5-026. John A. Roebling Suspension Bridge (formerly Covington-Cincinnati Suspension Bridge), Ohio River, between Cincinnati, Ohio, and Covington, Kentucky, 1856–1866. John A. Roebling, chief engineer. DETR, ca. 1907. P&P,LC-D4-70073.

The Civil War delayed construction of the 1,057-foot span. From 1895 to 1899, the original wood and iron deck was replaced by a stronger but heavier steel structure to support light rail traffic that required the addition of a second set of suspension cables. Wilhelm Hildenbrand (1843–1908), one of Roebling's senior engineers, designed the modification.

5-027. Approach to south portal, John A. Roebling Suspension Bridge (formerly Covington-Cincinnati Suspension Bridge), Ohio River, between Cincinnati, Ohio, and Covington, Kentucky, 1856–1866. John A. Roebling, chief engineer. Jack E. Boucher, photographer, 1982. P&P,HAER,OHIO,31-CINT, 45-5.

5-028

5-028. View of suspension cables and deck truss, John A. Roebling Suspension Bridge (formerly Covington-Cincinnati Suspension Bridge), Ohio River, between Cincinnati, Ohio, and Covington, Kentucky, 1856–1866. John A. Roebling, chief engineer. Jack E. Boucher, photographer, 1982. P&P,HAER,OHIO,31-CINT,45-7.

5-029. Sheet music cover for "Suspension Bridge Grand March" showing the John A. Roebling Suspension Bridge (formerly Covington-Cincinnati Suspension Bridge), Ohio River, between Cincinnati, Ohio, and Covington, Kentucky. Lithograph, published by C. Y. Fonda, Cincinnati, ca. 1867. P&P,LC-USZ62-7924.

5-029

5-030. Railroad Suspension Bridge, Great Central Railway, Niagara River, Niagara Falls, New York, 1855 (demolished). John A. Roebling, chief engineer. Currier & Ives, lithographer, 1856. P&P, LC-USZC2-3301.

Railroad suspension bridges are rare. Engineers have preferred the greater rigidity of truss spans when designing bridges for the loads imposed by rail traffic. This 821-foot bridge had wire cables and double wooden decks and stiffening trusses, which were replaced by iron in 1880. The lower deck served pedestrians and carriages.

5-031. Railroad advertisement showing Railroad Suspension Bridge, Great Central Railway, Niagara River, Niagara Falls, New York. Lithograph, ca. 1876. P&P,LC-USZC4-1304.

5-032. Railroad Suspension Bridge, Great Central Railway, Niagara River, Niagara Falls, New York, 1886 (demolished). Leffert L. Buck, chief engineer. Currier & Ives, lithograph, ca. 1886–1897. P&P, LC-USZC2-2879.

Leffert Buck (1837–1909) renovated Roebling's bridge to accommodate heavier trains by replacing the masonry towers with steel and rebuilding the deck. Ten years later, he replaced the renovated structure with a steel arch bridge (see 2-061).

5-033

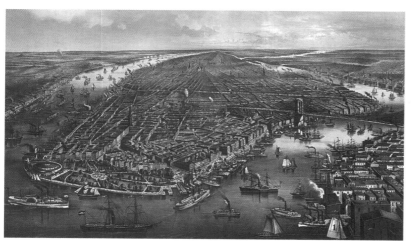

5-034

5-033. "The Great East River Suspension Bridge," lithograph of Brooklyn Bridge, East River, New York, New York, 1869–1883. John and Washington Roebling, chief engineers. Parsons and Atwater, delineators; Currier & Ives, publisher, ca. 1874. P&P,LC-USZC4-4632.

The Brooklyn Bridge has long ceded its title as longest suspension bridge but remains one of the most majestic structures in the United States and an enduring source of inspiration for engineers, architects, authors, and artists. Planned by John A. Roebling and completed by his son Washington, the 1,595-foot main span was more than 500 feet longer than Roebling's Cincinnati bridge (5-026), the previous record holder. Noteworthy features of its design and construction include the use of pneumatic caissons for the foundations, steel cables and trusses, the technique of spinning the cable bundles on site (based on an earlier spinning method by the French engineer Louis Vicat), and the diagonal stay cables that help stabilize the deflection of the deck under wind loads.

5-034. Bird's-eye view of New York. Ferdinand Meyer & Sons (New York, NY), lithographer; George Degen (New York, NY), publisher, 1873. P&P,PGA -Mayer (Ferdinand) & Sons—New York (C size).

This view anticipating the completion of the bridge shows it dwarfing every other structure in the city except the spire of Trinity Church. Such a view suggests the important role the bridge would play in the economic life of New York and Brooklyn.

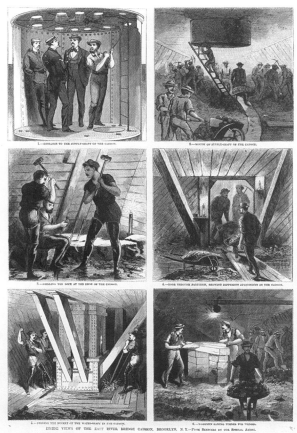

5-035

5-036

5-035. Demolition of buildings for the New York approach, Brooklyn Bridge, East River, New York, New York, 1869–1883. John and Washington Roebling, chief engineers. Wood engraving, W. P. Snyder, delineator, published in Harper's Weekly, November 24, 1877, p. 920. P&P,LC-USZ62-102100.

5-036. Workers in a caisson preparing the foundation for a tower, Brooklyn Bridge, East River, New York, New York, 1869–1883. John and Washington Roebling, chief engineers. Wood engraving, published in Frank Leslie's Illustrated Newspaper, October 15, 1870. P&P,LC-USZ62-124944.

Within months of each other in 1870, James Eads in St. Louis (see 2-050) followed by the Roeblings in New York employed pneumatic caissons, developed in France, for the construction of towers in the deep and turbulent waters of the Mississippi and the East rivers. A pneumatic caisson is an open-bottomed chamber that sinks as the tower is built upon its roof (see 3-285). Once embedded on the river floor, the chamber is pressurized with air to create a workspace for excavators, who enter via an airlock from a shaft leading to the surface. As the excavators remove mud and silt, the weight of the tower continues to sink the caisson down to bedrock. The effects of working in an atmosphere of high pressure were poorly understood at the time, and workers at both sites, including Washington Roebling, suffered from caisson disease, more commonly known as the bends.

5-037. Workers lashing stays to suspenders, Brooklyn Bridge, East River, New York, New York, 1869–1883. John and Washington Roebling, chief engineers. Wood engraving published in Frank Leslie's Illustrated Newspaper, April 28, 1883. P&P,LC-USZ62-102102.

5-037

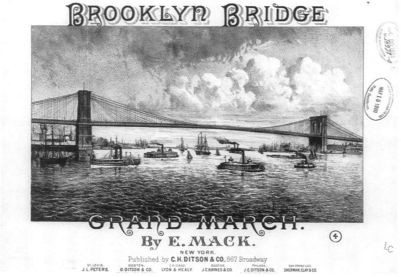

5-038. "Brooklyn Bridge Grand March," E. Mack, composer; C. H. Ditson (New York, NY), publisher, 1883. Music Division, M2.3.U6 A44.

5-039. Brooklyn Bridge, East River, New York, New York, 1869–1883. John and Washington Roebling, chief engineers. Jet T. Lowe, photographer, 1982. P&P,HAER,NY,31-NEYO,90-14.

5-040. Brooklyn approach, Brooklyn Bridge, East River, New York, New York, 1869–1883. John and Washington Roebling, chief engineers. Jet T. Lowe, photographer, 1982. P&P,HAER,NY,31-NEYO, 90-64.

5-041. Brooklyn tower, Brooklyn Bridge, East River, New York, New York, 1869–1883. John and Washington Roebling, chief engineers. Jet T. Lowe, photographer, 1982. P&P,HAER,NY,31-NEYO,90-25.

5-042. Cables and stays at tower cornice, Brooklyn Bridge, East River, New York, New York, 1869–1883. John and Washington Roebling, chief engineers. Jet T. Lowe, photographer, 1982. P&P,HAER,NY,31-NEYO,90-39.

5-043. Suspender connection at main cable, Brooklyn Bridge, East River, New York, New York, 1869–1883. John and Washington Roebling, chief engineers. Jet T. Lowe, photographer, 1982. P&P,HAER,NY,31-NEYO,90-45.

5-044

5-047

5-045

5-046

5-044. Cast-iron cable saddle inside Brooklyn tower, Brooklyn Bridge, East River, New York, New York, 1869–1883. John and Washington Roebling, chief engineers. Jet T. Lowe, photographer, 1982. P&P,HAER,NY,31-NEYO,90-40.

5-045. Suspender cables, diagonal stays, and deck-stiffening trusses (added by David Steinman ca. 1954), Brooklyn Bridge, East River, New York, New York, 1869–83. John and Washington Roebling, chief engineers. Jet T. Lowe, photographer, 1982. P&P,HAER,NY,31-NEYO,90-46.

5-046. Connection of main cable and deck at mid-span, Brooklyn Bridge, East River, New York, New York, 1869–1883. John and Washington Roebling, chief engineers. Jet T. Lowe, photographer, 1982. P&P,HAER,NY,31-NEYO,90-49.

5-047. Promenade, Brooklyn Bridge, East River, New York, New York, 1869–1883. John and Washington Roebling, chief engineers. DETR, ca. 1905–1920. P&P,LC-D4-36776.

5-048

5-049

SUSPENSION BRIDGES
ACROSS THE UNITED STATES

5-048. Bidwell Bar Suspension Bridge, Lake Oroville, Oroville Vic., Butte County, California, 1856 (originally spanned the Feather River, moved to present location ca. 1970). Jones & Murray (Sacramento, CA), builder; Starbuck Iron Works (Troy, NY), fabricator. Jet T. Lowe, photographer, 1984. P&P,HAER,CAL,4-ORO.V,1-2.

This was among the first suspension bridges erected in California. Its towers and cables were fabricated in upstate New York and shipped around Cape Horn.

5-049. Toll house, Bidwell Bar Suspension Bridge, Lake Oroville, Oroville Vic., Butte County, California, 1856 (originally spanned the Feather River, moved ca. 1970). Jet T. Lowe, photographer, 1984. P&P,HAER,CAL,4-ORO.V,1-3.

5-050. General William Dean Suspension Bridge, Kaskaskia River, Carlyle, Clinton County, Illinois, 1859–1861. Griffith D. Smith (PA), builder. Clark Bullard, photographer, 1936. P&P,HABS,ILL,14-CARL,1-2.

5-051. Elevation and plan, General William Dean Suspension Bridge, Kaskaskia River, Carlyle, Clinton County, Illinois, 1859–1861. Griffith D. Smith (PA), builder. Leslie A. Arons, delineator, 1936. P&P,HABS,ILL,14-CARL,1-,sheet no. 1.

5-050

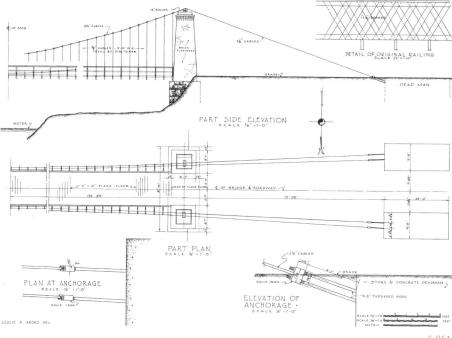

5-051

5-052

5-053

5-054

5-055

5-052. Deer Creek suspension bridge, Nevada City, Nevada County, California, 1861. Andrew S. Hallidie, builder. Photograph published by Lawrence & Houseworth, 1866. P&P,LC-USZ62-26929.

Best known as the inventor of the cable railroad, which he introduced in San Francisco in 1873, Andrew Hallidie (1836–1900) was a wire rope manufacturer who also applied his product to suspension bridges in the California gold fields (5-053).

5-053. Suspension bridge at Rattlesnake Bar, American River, Placer County, California, 1862. Andrew S. Hallidie, builder. Photograph published by Lawrence & Houseworth, 1866. P&P,LC-USZ62-26934.

5-054. New Portland suspension bridge, Carrabassett River, New Portland, Somerset County, Maine, 1866. David Elder, Charles B. Clark, builders. Jet T. Lowe, photographer, 1984. P&P,HAER,ME,13-NEWPO,1-1.

5-055. Detail of tower and anchorage, New Portland suspension bridge, Carrabassett River, New Portland, Somerset County, Maine, 1866. David Elder, Charles B. Clark, builders. Jet T. Lowe, photographer, 1984. P&P,HAER,ME,13-NEWPO,1-6.

5-056. View of tower, deck cables, and suspenders, New Portland suspension bridge, Carrabassett River, New Portland, Somerset County, Maine, 1866 (rehabilitated 1960). David Elder, Charles B. Clark, builders. Jet T. Lowe, photographer, 1984. P&P,HAER,ME,13-NEWPO,1-10.

5-057. Waco Suspension Bridge, Brazos River, Waco, McLennan County, Texas, 1868–1870, (reconstructed 1913–1914; rehabilitated as a pedestrian bridge 1976). Thomas M. Griffith, engineer; John A. Roebling's Sons Company (Trenton, NJ), cable fabricator; Missouri Valley Bridge & Iron Company, reconstruction. Jet T. Lowe, photographer, 1987. P&P,HAER,TEX,155-WACO,1-11.

Roebling's company supplied the cables and advised Griffith, based in New York City, on the design of this 475-foot bridge. The renovation of 1913 involved rebuilding the towers, replacing and reconfiguring the cables, and adding steel trusses.

5-058. Panoramic map, "Waco, Texas, County Seat of McLennan Cy." Beck & Pauli (Milwaukee, WI), lithographers, 1886. G&M, G4034.W2A3 1886 .N6.

As the first Brazos River crossing accessible to wagons, the Waco Suspension Bridge drew settlers and cattle drovers on the Chisholm Trail, who paid five cents a head to drive their herds across the span. The traffic attracted businesses, and the city grew rapidly in the decade following the bridge's completion. Waco's importance as a market and transportation hub received another boost in 1872 with the arrival of railroad lines, which crossed the Brazos on truss bridges.

5-059. Toll house, Waco Suspension Bridge, Brazos River, Waco, McLennan County, Texas, 1868–1870 (reconstructed, 1913–1914). Jet T. Lowe, photographer, 1987. P&P,HAER,TEX,155-WACO,1-20.

5-056

5-057

5-059

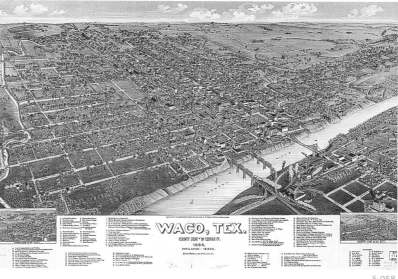

5-058

5-060

5-060. Boston Public Garden suspension bridge, Boston, Massachusetts, 1869. William G. Preston, architect. Martin Stupich, photographer, 1987. P&P,HAER,MASS,13-BOST,128-1.

Like the contemporary arch-shaped iron bridges of New York City's Central Park (1-019–1-020, 2-047–2-049), this design was driven primarily by aesthetic considerations.

5-061. Pedestrian suspension bridge, Philadelphia & Reading Railroad, Reading, Berks County, 1886–1887 (demolished 1983). Wilhelm Hildenbrand, engineer; John A. Roebling's Sons Company (Trenton, NJ), builder. William Edmund Barrett, photographer, 1975. P&P,HAER,PA,6-READ,12-1.

5-061

5-062. Lower Bridge at English Center, Little Pine Creek, English Center, Lycoming County, Pennsylvania, 1891. Dean & Westbrook (New York, NY), engineer, design and construction. Joseph Elliott, photographer, 1997. P&P,HAER,PA,41-ENGCE,1-2.

The 300-foot clear span is achieved by an unusual structure combining a pinned-eyebar chain with an arrangement of vertical and diagonal iron members forming a truss. The designers' intentions are unknown. Recent analysis indicates that the structure behaves as a trussed, inverted arch, but during construction it may have been thought of as a deck-stiffened suspension bridge.

5-063. Interior view, Lower Bridge at English Center, Little Pine Creek, English Center, Lycoming County, Pennsylvania, 1891. Dean & Westbrook (New York, NY), engineer, design and construction. Joseph Elliott, photographer, 1997. P&P,HAER,PA,41-ENGCE,1-6.

5-064. Williamsburg Bridge, East River, New York, New York, 1896–1903. Leffert L. Buck, chief engineer. DETR, ca. 1902–1910. P&P,LC-D4-33447.

With a span of 1,600 feet, the Williamsburg Bridge is slightly longer than the nearby Brooklyn Bridge, and its deck, stiffened by deep trusses to accommodate rail traffic, is 40 feet wider. For the first time on a major bridge, the towers are steel.

5-065

5-066

Twentieth-Century Suspension Bridges

5-065. Essex-Merrimac Bridge (Chain Bridge), Merrimack River, Newburyport, Essex County, Massachusetts, 1909–1910. George F. Swain, Robert R. Evans, engineers; Holbrook, Cabot & Rollins Corporation (Boston, MA), builder. Martin Stupich, photographer, 1990. P&P,HAER,MASS,5-AMB,3-2.

The design of this bridge recalls the early-nineteenth-century suspension bridges that preceded it on the site.

5-066. Manhattan Bridge, East River, New York, New York, 1901–1909. R. S. Buck, Gustav Lindenthal, Othniel Foster Nichols, Ralph Modjeski, Leon Moisseiff, engineers; Carrère & Hastings, consulting architects; John A. Roebling's Sons Company (Trenton, NJ), steel cable. Jet T. Lowe, photographer, 1979. P&P,HAER,NY,31-NEYO,164-7.

Municipal politics and fierce professional rivalries impinged upon the design process and the tenure of the successive engineers charged with construction of this 1,470-foot span across the East River. Although it was initially conceived by Buck as a cable-supported structure, Lindenthal redesigned it by replacing the cables with steel eyebars—only to be eclipsed by Nichols and Modjeski, who imposed their own vision using cables. They based their design on the deflection theory developed by Josef Melan in 1888 and introduced to the United States by Leon Moisseiff. Widely used by engineers in the first half of the twentieth century, it provided a means of analyzing the primary stresses of the suspended span.

5-067. View of deck construction, Manhattan Bridge, East River, New York, New York, 1901–1909. R. S. Buck, Gustav Lindenthal, Othniel Foster Nichols, Ralph Modjeski, Leon Moisseiff, engineers; Carrère & Hastings, consulting architects; John A. Roebling's Sons Company (Trenton, NJ), steel cable. Irving Underhill, photographer, 1909. P&P,LC-USZ62-100104.

5-068. Stereo view of workers attaching suspender cables, Manhattan Bridge, East River, New York, New York, 1901–1909. R. S. Buck, Gustav Lindenthal, Othniel Foster Nichols, Ralph Modjeski, Leon Moisseiff, engineers; Carrère & Hastings, consulting architects; John A. Roebling's Sons Company (Trenton, NJ), steel cable. H. C. White Company, publisher, 1909. P&P,LC-USZ62-79052.

5-069

5-069. Dresden Suspension Bridge, Muskingum River, Dresden, Muskingum County, Ohio, 1914. Bellefontaine Bridge & Steel Company (Bellefontaine, OH), builder. Joseph Elliott, photographer, 1992. P&P,HAER,OHIO,60-DRES,1-4.

This bridge has a suspended span of 705 feet and employs eyebars such as those Lindenthal had proposed for the Manhattan Bridge.

5-070. Detail of eyebars at anchorage, Dresden Suspension Bridge, Muskingum River, Dresden, Muskingum County, Ohio, 1914. Bellefontaine Bridge & Steel Company (Bellefontaine, OH), builder. Joseph Elliott, photographer, 1992. P&P,HAER,OHIO,60-DRES,1-7.

5-070

5-071. Winkley Bridge, Little Red River, Heber Springs, Cleburne County, Arkansas, 1912 (collapsed 1989). Harry Churchill (Pangburn, AR), builder. Louise T. Cawood, photographer, 1988. P&P,HAER,ARK,12-HESP,1-2.

5-072. Dewey Suspension Bridge, Colorado River, Cisco Vic., Grand County, Utah, 1915–1916. Midland Bridge Company (Kansas City, MO), builder. Brian Grogan, photographer, 1993. P&P,HAER,UTAH,10-CIS,V.1-1.

5-073. Mid-Hudson Bridge, Hudson River, Poughkeepsie, Dutchess County, New York, 1925–1930. Ralph Modjeski, Daniel Moran, engineers. Jet T. Lowe, photographer, 1986. P&P,HAER,NY,14-POKEP,7-4.

5-071

5-072

5-073

5-074

5-075

5-074. San Rafael Bridge, San Rafael River, Castle Dale Vic., Emery County, Utah, 1937. Civilian Conservation Corps, builder. John Pendlebury, photographer, 1989. P&P,HAER,UTAH,8-CADA.V,1-1.

5-075. Detail of suspension towers and wood Howe stiffening truss, San Rafael Bridge, San Rafael River, Castle Dale Vic., Emery County, Utah, 1937. Civilian Conservation Corps, builder. John Pendlebury, photographer, 1989. P&P,HAER,UTAH,8-CADA.V,1-7.

5-076

5-077

5-076. Sheep drive on Verde River Sheep Bridge, Cave Creek Vic., Yavapai County, Arizona, 1943 (replaced 1989). Cyril O. Gillam, engineer. Unidentified photographer, ca. 1944. P&P,HAER,ARIZ,13-CACR.V,1-39.

Designed by an engineer but built by ranch hands with locally available materials, including cables previously utilized by mining operations, this and similar suspension bridges served the permanent driveways established at the beginning of the twentieth century for herding sheep between winter and summer ranges.

5-077. Verde River Sheep Bridge, Cave Creek Vic., Yavapai County, Arizona, 1943 (replaced 1989). Cyril O. Gillam, engineer. Dale A. Politi, photographer, 1987. P&P,HAER,ARIZ,13-CACR.V,1-9.

5-078. Frank Auza, rancher and builder of Verde River Sheep Bridge, Cave Creek Vic., Yavapai County, Arizona, 1943 (replaced 1989). Cyril O. Gillam, engineer. Dale A. Politi, photographer, 1987. P&P,HAER,ARIZ,13-CACR.V,1-4.

5-078

5-079

5-080

5-081

5-079. Beaver Bridge, White River, Beaver, Carroll County, Arkansas, 1949. Pioneer Construction Company (Malvern, AR), builder. Jeff Holder and Michael Swanda, photographers, 1988. P&P,HAER,ARK,8-BEAV,1-2.

5-080. Chow Chow Suspension Bridge, Quinhault River, Quinhault Indian Reservation, Taholah Vic., Grays Harbor County, Washington, ca. 1950 (demolished). Frank Millward, designer. Jet T. Lowe, photographer, 1979. P&P,HAER,WASH,14-TAH.V,1-1.

5-081. Detail of pylons and king-post stiffening trusses, Chow Chow Suspension Bridge, Quinhault River, Quinhault Indian Reservation, Taholah Vic., Grays Harbor County, Washington, ca. 1950 (demolished). Frank Millward, designer. Jet T. Lowe, photographer, 1979. P&P,HAER,WASH,14-TAH.V,1-4.

TACOMA NARROWS SUSPENSION BRIDGES

5-082. Tacoma Narrows Bridge, Puget Sound, Tacoma, Pierce County, Washington, 1938–1940. Leon Moisseiff, engineer, superstructure; Clark H. Eldridge, engineer, piers; Pacific Bridge Company (Portland, OR), Bethlehem Steel Company (Bethlehem, PA), builders. Unidentified photographer, 1940, from Clark H. Eldridge, "Tacoma Narrows Bridge, Tacoma, Washington, Final Report on Design and Construction," 1941. P&P, HAER,WASH, 27-TACO,11-31.

Only four months after its opening, the slender 2,800-foot main span broke apart after gale-strength winds generated oscillations of 28 feet (see IN-022). The failure discredited the popular deflection theory that had guided the structural analysis of suspension bridges since the early twentieth century.

5-083. Tacoma Narrows Bridge, Puget Sound, Tacoma, Pierce County, Washington, 1948–1950). Charles E. Andrew, chief engineer; Dexter R. Smith, design engineer; Bethlehem Pacific Coast Steel Corporation (San Francisco, CA), John A. Roebling's Sons Company (Trenton, NJ), builders. Jet T. Lowe, photographer, 1993. P&P, HAER,WASH,27-TACO,11-19.

For the design of the replacement bridge, wind tunnel tests were used for the first time to study the behavior of the new suspension system, which featured larger towers, cable stays, and deeper stiffening trusses.

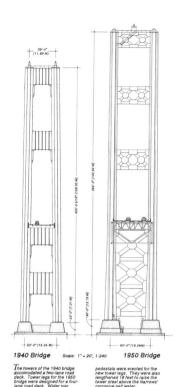

Tower Elevations

39'-0" (11.89 M)

430'-0 5/16" (128.02 M)

8'-0 1/2" (2.45 M)

50'-0" (15.24 M)

1940 Bridge   Scale: 1" = 20', 1:240

The towers of the 1940 bridge accomodated a two-lane road deck. Tower legs for the 1950 bridge were designed for a four-lane road deck. Wider pier

460'-0 5/16" (140.82 M)

23'-0" (7.01 M)

60'-0" (18.29M)

1950 Bridge

pedestals were erected for the new tower legs. They were also lengthened 18 feet to raise the tower steel above the Narrows' corrosive salt water.

5-084

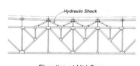

21'-8" (6.55 M)

33'-0" (10.06 M)

1940 Bridge Roadway

Axonometrics Scale: 1/8" = 1'-0", 1:96

The Tacoma Narrows Bridge collapsed on 7 November 1940 during a gale between 35 to 42 miles per hour, with a wind pressure of only five pounds per square foot. The steady wind's effects on the structure produced a fluctuating resultant force that synchronized in timing and direction with the bridge's natural harmonic motions (figs. 1&2), progressively amplifying them to destructive levels. Both vertical and torsional oscillations contributed to the failure of the bridge. The bridge's inherent weakness and susceptibility to these winds lay in its shallow stiffening girders and its narrow roadway.

Theodore von Karman, who had pioneered wind tunnel analysis at the California Institute of Technology, argued that the bridge deck's aerodynamic shape was a more important factor in its failure than its lightness and flexibility. Von Karman suspected that the bridge had experienced vortex shedding, a condition where objects like airplane wings or bridge decks displace air flowing around them and form eddies or vortices, which may induce vibration in the object (figs 3&4). He believed that wind flowing over the bridge's solid girder side plates created shedding that when

Damping Mechanism
1950 Bridge
Scale: 1/16" = 1'-0", 1:192

Hydraulic Shock

Hydraulic Shock

Elevation at Towers

30'-0" (9.14 M)

1950 Bridge Roadway

combined with the flutter and resonance already present in the deck produced the violent oscillations that caused the catastrophic failure (figs. 5&6).

Designing the replacement bridge's deck stiffening system involved subjecting dynamic scale models to wind tunnel testing to better understand wind effects on them.

Designers for the 1950 bridge were not satisfied with their ability to eliminate torsional and vertical movements in their proposed structure. They hoped to enhance their design's natural damping ability with mechanical devices. One of these was a double-lateral bracing system in the stiffening truss. It increased torsional frequency motion and tortional stiffness.

Hydraulic Shock

Elevation at Mid-Span

To eliminate torsional and vertical movement cylindrical hydraulic shock absorbers were used at three points on the bridge: coupling the top of the stiffening truss at mid-span with the suspension cables, connecting between the top chords of the main span and side span stiffening trusses, and extending as outriggers from the trusses' bottom chords to the towers.

1940 Bridge Failure
Diagrams & Illustrations

FIGURE 1

FIGURE 2

Lift

Drag

Wind

FIGURE 3

FIGURE 4

FIGURE 5

FIGURE 6

5-084. Comparison of towers, structural details, Tacoma Narrows Bridges, 1940, 1950. Karl W. Stumpf, Wolfgang G. Mayr, delineators, 1993. P&P,HAER,WASH,27-TACO,11-,sheet no. 2.

SUSPENSION BRIDGES BY DAVID B. STEINMAN AND HOLTON D. ROBINSON

5-085. St. John's Bridge, Willamette River, Portland, Oregon, 1929–1931. David B. Steinman, Robinson & Steinman, engineers; Wallace Bridge & Structural Steel Company (Seattle, WA), John A. Roebling's Sons Company (Trenton, NJ), fabricators. Jet T. Lowe, photographer, 1990. P&P,HAER,ORE,26-PORT,13-1.

David B. Steinman (1886—1960) and his partner Holton D. Robinson (1863—1945) established themselves as leading designers of bridges in the 1920s and remained at the pinnacle of their profession until their deaths. Steinman, indeed, achieved prominence among a broader public as the author of books on bridges and on John Roebling, and as a poet. With this bridge, which has a main span of 1,207 feet, Steinman and Robinson introduced the use of prestressed, twisted-wire cables that eliminated the task of spinning wires in situ, which had been a time-consuming aspect of wire-cable construction for long-span structures.

5-085

5-086. Waldo-Hancock Bridge, Penobscot River, Bucksport Vic., Hancock County, Maine, 1929–1931 (demolished). David B. Steinman, Robinson & Steinman, designer; American Bridge Company (Pittsburgh, PA), superstructure; John A. Roebling Sons Company (Trenton, NJ), cables; Merritt-Chapman & Scott Corporation (New York, NY) substructure, fabricators. Jet T. Lowe, photographer, 1994. P&P,HAER,ME,5-BUCK.V,1-3.

Designed at the same time as St. John's Bridge in Portland, Oregon, this bridge also utilized prestressed, twisted-wire cables. Its towers are noteworthy as the first to be designed as Vierendeel trusses, which would later be used for the towers of the Golden Gate (5-095) and Triborough (5-102) bridges. (Named after Belgian engineer Arthur Vierendeel (1852–1940), Vierendeel trusses are composed of rectangular or trapezoidal panels and rigid connections.) Corrosion of the cables necessitated construction of a new bridge in 2006.

5-087. Deer Isle–Sedgwick Bridge, Eggemoggin Reach, Sedgwick, Hancock County, Maine, 1937–1939. David B. Steinman, Robinson & Steinman, chief engineer; Phoenix Bridge Company (Phoenixville, PA), superstructure; Merritt-Chapman & Scott (New York, NY), substructure. Jet T. Lowe, photographer, 1994. P&P,HAER,ME,5-SEDG.V,1-3.

A tight construction schedule and difficult working conditions required Steinman and Robinson to develop techniques for prefabrication and assembly that simplified on-site adjustments. Examples of such innovations included prefabricated cofferdams (a type of caisson) brought to the site on barges and the use of adjustable connectors for mooring the main cable strands.

5-086

5-087

5-088

5-091

5-088. Mackinac Bridge, Mackinac Straits, Mackinaw City, Cheboygan County, Michigan, 1953–1957. David Steinman, chief engineer; Merritt-Chapman & Scott (New York, NY), substructure; American Bridge Company (Pittsburgh, PA), superstructure. Unidentified photographer, 1958. P&P, NYWTS.

Steinman regarded this bridge, the last of his design to be completed during his lifetime, as his crowning achievement. At the time of its dedication, its total suspended span of 8,614 feet from anchorage to anchorage was the longest in the world. Its deep stiffening trusses and open, metal-grate roadway in the center lanes of the main span are Steinman's responses to the lessons regarding wind-induced oscillations learned from the Tacoma Narrows Bridge disaster in 1940 (see IN-024, 5-082).

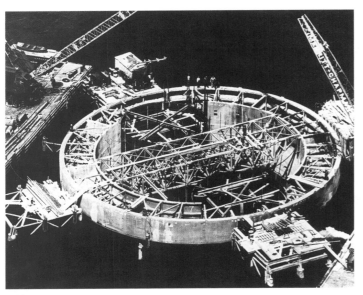

5-089

5-089. Tower foundation, Mackinac Bridge, Mackinac Straits, Mackinaw City, Cheboygan County, Michigan, 1953–1957. David Steinman, chief engineer; Merritt-Chapman & Scott (New York, NY), substructure; American Bridge Company (Pittsburgh, PA), superstructure. United Press Photo, June 1955. P&P, NYWTS.

5-090. Prefabricated truss being moved into position by barge, Mackinac Bridge, Mackinac Straits, Mackinaw City, Cheboygan County, Michigan, 1953–1957. David Steinman, chief engineer; Merritt-Chapman & Scott (New York, NY), substructure; American Bridge Company (Pittsburgh, PA), superstructure. United Press Photo, November 1955. P&P, NYWTS.

5-091. Construction photo showing suspension span before installation of deck, Mackinac Bridge, Mackinac Straits, Mackinaw City, Cheboygan County, Michigan, 1953–1957. David Steinman, chief engineer; Merritt-Chapman & Scott (New York, NY), substructure; American Bridge Company (Pittsburgh, PA), superstructure. United Press Photo, December 1956. P&P, NYWTS.

5-090

5-092. View from Yerba Buena Island, West Bay crossing, San Francisco–Oakland Bay Bridge, San Francisco Bay, between San Francisco and Oakland, California, 1933–1936. C. H. Purcell, chief engineer; Columbia Steel Company (Portland, OR) and American Bridge Company (Pittsburgh, PA), superstructure; Transbay Construction Company, substructure. Jet T. Lowe, photographer, 1985. P&P,HAER,CAL,38-SANFRA,141-4.

The western crossing of the San Francisco–Oakland Bay Bridge (see 3-379 for the eastern crossing) consists of two 2,310-foot suspension bridges, which share a common anchorage in the middle of the bay, a unique arrangement in the United States.

5-093. Aerial view, West Bay crossing, San Francisco–Oakland Bay Bridge, San Francisco Bay, between San Francisco and Oakland, California, 1933–1936. C. H. Purcell, chief engineer; Columbia Steel Company (Portland, OR) and American Bridge Company (Pittsburgh, PA), superstructure; Transbay Construction Company, substructure. Pigett Photo, ca. 1936. P&P,LC-USZ62-127223.

5-094. Aerial view of the Pan American Airways Philippine Martin Clipper flying over the West Bay crossing of the San Francisco–Oakland Bay Bridge. Clyde H. Sunderland, photographer, ca. 1936. P&P,LC-USZ62-45025.

5-092

5-093

5-094

5-095

5-096. Aerial view of construction, Golden Gate Bridge, San Francisco Bay, San Francisco, California, 1933–1937. Joseph B. Strauss, chief engineer; O. H. Ammann, Charles Derleth Jr., Charles Ellis, Leon S. Moisseiff, consulting engineers; Irving F. and Gertrude C. Morrow, consulting architects; Bethlehem Steel Company (Bethlehem, PA), superstructure; John A. Roebling's Sons Company (Trenton, NJ), cables; Pacific Bridge Company (Portland, OR), substructure. Charles M. Hiller, photographer, ca. 1934. P&P,LC-USZ62-100678.

5-095. Golden Gate Bridge, San Francisco Bay, San Francisco, California, 1933–1937. Joseph B. Strauss, chief engineer; O. H. Ammann, Charles Derleth Jr., Charles Ellis, Leon S. Moisseiff, consulting engineers; Irving F. and Gertrude C. Morrow, consulting architects; Bethlehem Steel Company (Bethlehem, PA), superstructure; John A. Roebling's Sons Company (Trenton, NJ), cables; Pacific Bridge Company (Portland, OR), substructure. Jet T. Lowe, photographer, 1984. P&P,HAER,CAL,38-SANFRA,140-4.

One of the world's most photogenic bridges, the Golden Gate Bridge is also a spectacular engineering achievement because of the length of the main span (4,200 feet) and the construction challenges posed by the site's strong winds, currents, and fickle weather. Offering motorists a shortcut across San Francisco Bay, the bridge played an important role in the development of Marin County.

5-096

5-097. View showing steel arch of San Francisco approach span, Golden Gate Bridge, San Francisco Bay, San Francisco, California, 1933–1937. Joseph B. Strauss, chief engineer; O. H. Ammann, Charles Derleth Jr., Charles Ellis, Leon S. Moisseiff, consulting engineers; Irving F. and Gertrude C. Morrow, consulting architects; Bethlehem Steel Company (Bethlehem, PA), superstructure; John A. Roebling's Sons Company (Trenton, NJ), cables; Pacific Bridge Company (Portland, OR), substructure. Jet T. Lowe, photographer, 1984. P&P, HAER,CAL,38-SANFRA,140-13.

5-098. Tower, Golden Gate Bridge, San Francisco Bay, San Francisco, California, 1933–1937. Joseph B. Strauss, chief engineer; O. H. Ammann, Charles Derleth Jr., Charles Ellis, Leon S. Moisseiff, consulting engineers; Irving F. and Gertrude C. Morrow, consulting architects; Bethlehem Steel Company (Bethlehem, PA), superstructure; John A. Roebling's Sons Company (Trenton, NJ), cables; Pacific Bridge Company (Portland, OR), substructure. Jet T. Lowe, photographer, 1984. P&P,HAER,CAL,38-SANFRA,140-12.

Strauss and the consulting architects designed the towers as Vierendeel trusses.

5-099. Cable saddle on south tower, Golden Gate Bridge, San Francisco Bay, San Francisco, California, 1933–1937. Joseph B. Strauss, chief engineer; O. H. Ammann, Charles Derleth Jr., Charles Ellis, Leon S. Moisseiff, consulting engineers; Irving F. and Gertrude C. Morrow, consulting architects, Bethlehem Steel Company (Bethlehem, PA), superstructure; John A. Roebling's Sons Company (Trenton, NJ), cables; Pacific Bridge Company (Portland, OR), substructure. Jet T. Lowe, photographer, 1984. P&P,HAER,CAL,38-SANFRA,140-31.

5-100

5-100. View from south tower along main cable, Golden Gate Bridge, San Francisco Bay, San Francisco, California, 1933–1937. Joseph B. Strauss, chief engineer; O. H. Ammann, Charles Derleth Jr., Charles Ellis, Leon S. Moisseiff, consulting engineers; Irving F. and Gertrude C. Morrow, consulting architects; Bethlehem Steel Company (Bethlehem, PA), superstructure; John A. Roebling's Sons Company (Trenton, NJ), cables; Pacific Bridge Company (Portland, OR), substructure. Jet T. Lowe, photographer, 1984. P&P,HAER,CAL,38-SAN-FRA,140-28.

5-101. Connections of main cable, suspender, and deck beam, Golden Gate Bridge, San Francisco Bay, San Francisco, California, 1933–1937. Joseph B. Strauss, chief engineer; O. H. Ammann, Charles Derleth Jr., Charles Ellis, Leon S. Moisseiff, consulting engineers; Irving F. and Gertrude C. Morrow, consulting architects; Bethlehem Steel Company (Bethlehem, PA), superstructure; John A. Roebling's Sons Company (Trenton, NJ), cables; Pacific Bridge Company (Portland, OR), substructure. Jet T. Lowe, photographer, 1984. P&P,HAER, CAL,38-SANFRA,140-34.

5-101

SUSPENSION BRIDGES BY OTHMAR AMMANN AND ALLSTON DANA

5-102. Triborough Bridge, East River at Hell Gate, New York, New York, 1929–1936. Othmar Ammann, chief engineer; Allston Dana, design engineer; Aymar Embury II, consulting architect. Jet T. Lowe, photographer, 1991. P&P,HAER,NY,41-QUE,2-4.

The 1,380-foot suspension span designed shortly after Ammann and Dana had completed the George Washington Bridge (5-006) is part of a network consisting of three bridges, a viaduct, and 14 miles of roads linking the boroughs of Manhattan, Queens, and the Bronx. Gustav Lindenthal, engineer of the nearby Hell Gate Bridge (see 2-072), worried that the "cheap pole and washline architecture" of the new bridge would diminish the appearance of his steel arch.

5-103. Exchange plaza on Randall's Island, Triborough Bridge, East River at Hell Gate, New York, New York, 1933–1936. Jet T. Lowe, photographer, 1991. P&P,HAER,NY,41-QUE,2-21.

5-104. Bronx-Whitestone Bridge, East River, New York, New York, 1937–1939. Othmar Ammann, chief engineer; Allston Dana, design engineer; Aymar Embury II, consulting architect. Jet T. Lowe, photographer, 1991. P&P,HAER,NY,3-BRONX,14-10.

Ammann and Dana designed the bridge to be stiffened by slender plate girders. Following the collapse of the Tacoma Bay Bridge in 1940, they added diagonal stays and 14-foot-deep Warren trusses to provide additional resistance to extreme oscillations induced by wind. The trusses were removed in 2004 as part of an extensive rehabilitation utilizing lighter devices for managing aerodynamic stresses.

5-102

5-103

5-104

5-105

5-106

5-105. Verrazano-Narrows Bridge, New York, New York, 1959–1964. Othmar Ammann, chief engineer; Ammann & Whitney, consulting engineers; American Bridge Company (Pittsburgh, PA), cables and deck. Jet T. Lowe, photographer, 1991.
P&P,HAER,NY,24-BROK,57-15.

The last bridge designed under Ammann's direction, the Verrazano has a main span of 4,260 feet, the longest in the world until 1981, when the Humber Bridge in England surpassed it.

5-106. Lower deck, Verrazano-Narrows Bridge, New York, New York, 1959–1964. Othmar Ammann, chief engineer; Ammann & Whitney, consulting engineers; American Bridge Company (Pittsburgh, PA), cables and deck. Jet T. Lowe, photographer, 1991.
P&P,HAER,NY,24-BROK,57-22.

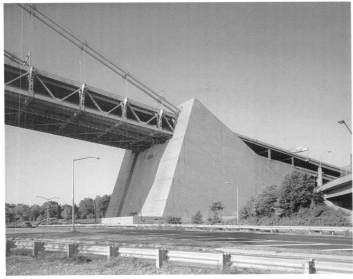

5-107. Anchorage, Verrazano-Narrows Bridge, New York, New York, 1959–1964. Othmar Ammann, chief engineer; Ammann & Whitney, consulting engineers; American Bridge Company (Pittsburgh, PA), cables and deck. Jet T. Lowe, photographer, 1991. P&P,HAER,NY,24-BROK,57-17.

5-108. Interior of anchorage, Verrazano-Narrows Bridge, New York, New York, 1959–1964. Othmar Ammann, chief engineer; Ammann & Whitney, consulting engineers; American Bridge Company (Pittsburgh, PA), cables and deck. Jet T. Lowe, photographer, 1991. P&P,HAER,NY,24-BROK,57-28.

5-109. Cable strand anchors, Verrazano-Narrows Bridge, New York, New York, 1959–1964. Othmar Ammann, chief engineer; Ammann & Whitney, consulting engineers; American Bridge Company (Pittsburgh, PA), cables and deck. Jet T. Lowe, photographer, 1991. P&P,HAER,NY,24-BROK,57-30.

CABLE-STAYED BRIDGES

Cable-stayed bridges resemble suspension bridges in their use of cables to support the deck, but they are structurally distinct because the cables run directly from the deck to a tower. Although a number of cable-stayed designs were developed in the early nineteenth century, they were dismissed in the influential writings of Claude-Louis-Marie-Henri Navier. Some engineers, however, such as John A. Roebling, did incorporate diagonal stays in suspension bridges. Little research on pure cable-stayed designs was undertaken until the late 1940s, when Franz Dischinger (1887–1953) and other European engineers employed them as economical and aesthetically pleasing solutions for the replacement of medium- and long-span bridges destroyed during World War II. Apart from a few isolated examples, cable-stayed designs were not adopted by American engineers until the 1970s, but since that belated introduction, their use has increased rapidly. Their dramatic profiles have acquired an aura of modernity similar to that associated with the great suspension bridges of the mid-twentieth century.

5-110

5-110. Bluff Dale suspension bridge, Pauluxy River, Bluff Dale, Erath County, Texas, 1890 (moved upstream 1934). Edwin Elijah Runyon, Runyon Bridge Company (Weatherford, TX), builder. Joseph Elliott, photographer, 1996. P&P,HAER,TX,72-BLUDA,1-3.

A regional bridge builder, Runyon held patents for the design of this unconventional structure, which has a main span of 140 feet. It is among the oldest surviving cable-stayed bridges in the United States.

5-111. Detail of column, Bluff Dale suspension bridge, Pauluxy River, Bluff Dale, Erath County, Texas, 1890 (moved upstream 1934). Edwin Elijah Runyon, Runyon Bridge Company (Weatherford, TX), builder. Joseph Elliott, photographer, 1996. P&P,HAER,TX,72-BLUDA,1-7.

5-112. Cable connections at deck, Bluff Dale suspension bridge, Pauluxy River, Bluff Dale, Erath County, Texas, 1890 (moved upstream 1934). Edwin Elijah Runyon, Runyon Bridge Company (Weatherford, TX), builder. Joseph Elliott, photographer, 1996. P&P,HAER,TX,72-BLUDA,1-9.

5-113. Sitka Harbor Bridge (O'Connell Bridge), Sitka, Sitka Borough, Alaska, 1971–1972. Alaska Department of Highways, designer; Associated Engineers & Contractors, builder. Jet T. Lowe, photographer, 1991. P&P,HAER,AK,17-SITKA,11-1.

With a main span of 450 feet, this was the first modern cable-stayed bridge built in the United States.

5-111

5-112

5-113

5-114

5-114. Pasco-Kennewick Bridge, Columbia River, Pasco, Franklin County, Washington, 1978. Arvid Grant; Loenhardt, Andrä, and Partners, engineers. Unidentified photographer, ca. 1980. P&P,HAER,WASH,11-PASC,1-14.

This bridge has reinforced concrete towers and a main span of 981 feet.

5-115. Main span, Sunshine Skyway, Tampa Bay, Pinellas and Manatee Counties, Florida, 1978–1987. Figg & Muller Engineers, Inc., designers; Pashen Contractors (Chicago, IL), American Bridge Company (Chicago, IL), and Morrison-Knudson (Boise, ID), builders, high-level approaches and main span. P&P,LC-DIG-ppem-00058.

During a storm in 1980, a freighter rammed a pier and caused the collapse of 1,200 feet of cantilevered trusses, taking the lives of thirty-five people. The new bridge, designed to resist catastrophic damage from ships and hurricane winds, was the nation's largest cable-stay structure upon completion. Unlike the Sitka and Pasco-Kennewick bridges, its concrete deck is supported with a single plane of cables attached along the centerline. Equally innovative was the construction of the deck using precast box-girder segments measuring 12 feet long, 95 feet wide, and 15 feet deep.

5-116. Detail of main span, Clark Bridge, Mississippi River, Alton, Madison County, Illinois, 1990–1994. Hanson Professional Services Inc. and Figg Engineering Group, designers; McCarthy Brothers Company (St. Louis, MO) and PCL Construction (Denver, CO), builders, main span superstructure. David Plowden, photographer, 1998. P&P,LC-DIG-ppbd-00351.

The 1,360-foot main span is supported by two sets of cables each passing over slender, single towers and arranged to form two planes of stays attached to the edges of the deck.

5-115

5-116

5-117. Leonard P. Zakim Bunker Hill Bridge, Charles River, Boston, Massachusetts, 1997–2003. Christian Menn, HNTB Corporation, engineers; Miguel Rosales, architect and urban designer; Bechtel/Parsons Brinkerhoff, project managers; Atkinson-Kewit, contractors. Andy Ryan, photographer, 2005. © Andy Ryan, 2005. P&P,LC-DIG-ppbd-00349.

Part of Boston's Central Artery highway project, the bridge carries ten lanes of traffic. Its 183-foot width makes it the world's widest cable-stayed bridge. The main span is 749 feet long. Due to extremely tight constraints on the site including interstate highway ramps, connection to a tunnel, and adjacent urban neighborhoods, the configuration of steel and concrete composite towers and the cables is assymmetrical. The cables of the main span are in two planes supporting the deck at the edges; the back stays are in a single plane anchored along the center line.

5-118. Sundial Bridge, Sacramento River, Redding, Shasta County, California, 2004. Santiago Calatrava, designer; Kiewit Pacific Company (Vancouver, Canada), builder. Alan Karchmer, photographer, 2004. © ESTO 2004. P&P,LC-DIG-ppbd-00350.

This is the first bridge in the United States designed by the Spanish architect and engineer Santiago Calatrava. As the centerpiece of Turtle Bay Exploration Park, its design is intended to heighten awareness of the site. The cables supporting the 748-foot span are anchored by an inclined steel pylon aligned to function as a sundial, and the glass-and-steel deck allows views of the river, which is a spawning ground for salmon.

5-117

5-118

BIBLIOGRAPHY

GENERAL WORKS

Billington, David P. *The Tower and the Bridge: The New Art of Structural Engineering.* New York: Basic Books, 1983.

Brainerd, Wesley. *Bridge Building in Wartime: Colonel Wesley Brainerd's Memoir of the 50th New York Volunteer Engineers.* Edited by Ed Malles. Knoxville: University of Tennessee Press, 1997.

Bucciarelli, Louis. "An Ethnographic Perspective on Engineering Design." *Design Studies* 8, no. 3 (July 1988): 159–68.

Building Arts Forum/New York. *Bridging the Gap: Rethinking the Relationship of Architect and Engineer.* Proceedings of the Building Arts Forum/New York symposium, Guggenheim Museum, April 1989. Deborah Gans, ed. New York: Van Nostrand Reinhold, 1991.

Condit, Carl W. *American Building Art: The Nineteenth Century.* New York: Oxford University Press, 1960.

_____. *American Building Art: The Twentieth Century.* New York: Oxford University Press, 1961.

Crane, Hart. *The Bridge: A Poem.* New York: Horace Liveright, 1930.

Danko, George. "The Evolution of the Simple Truss Bridge 1790 to 1850: From Empiricism to Scientific Construction." PhD diss., University of Pennsylvania, 1979.

Darnell, Victor. *A Directory of American Bridge-Building Companies, 1840–1900.* Washington, D.C.: Society for Industrial Archeology, 1984.

_____. "The Other Literature of Bridge Building," *IA, The Journal of the Society for Industrial Archeology* 15, no. 2 (1989): 40–56.

DeLony, Eric. "The Golden Age of the Iron Bridge." *Invention & Technology* 10, no. 2 (Fall 1994): 8–22.

_____. *Landmark American Bridges.* New York: American Society of Civil Engineers, 1993.

_____. "Surviving Cast- and Wrought-Iron Bridges in America." *IA, The Journal of the Society for Industrial Archeology* 19, no. 2 (1993): 17–47.

Haupt, Herman. *General Theory of Bridge Construction.* New York: D. Appleton & Co., 1851.

Jackson, Donald C. *Great American Bridges and Dams.* New York: John Wiley & Sons, 1988.

Kemp, Emory L., ed. *American Bridge Patents: The First Century (1790–1890).* Morgantown: West Virginia University Press, 2005.

_____. "The Fabric of Historic Bridges." *IA, The Journal of the Society for Industrial Archeology* 15, no. 2 (1989): 3–22.

_____. "The Introduction of Cast and Wrought Iron in Bridge Building." *IA, The Journal of the Society for Industrial Archeology* 19, no. 2 (1993): 5–16.

_____. "National Styles in Engineering: The Case of the 19th-Century Suspension Bridge." *IA, The Journal of the Society for Industrial Archeology* 19, no. 1 (1993): 21–36.

Kranakis, Eda. *Constructing a Bridge: An Exploration of Engineering Culture, Design, and Research in Nineteenth-Century France and America*. Cambridge, MA: MIT Press, 1997.

Leonhardt, Fritz. *Brücken / Bridges: Aesthetics and Design*. Cambridge, MA: MIT Press, 1984.

Mock [Kassler], Elizabeth B. *The Architecture of Bridges*. New York: Museum of Modern Art, 1949.

Petroski, Henry. *Engineers of Dreams: Great Bridge Builders and the Spanning of America*. New York: Knopf, 1995.

Plowden, David. *Bridges: The Spans of North America*. Rev. ed. New York: W. W. Norton, 2002. First published 1974.

Salvadori, Mario. *Why Buildings Stand Up: The Strength of Architecture*. New York: W. W. Norton, 1980.

Seely, Bruce E. *Building the American Highway System: Engineers as Policy Makers*. Philadelphia: Temple University Press, 1987.

Steinman, David B. *I Built a Bridge, and Other Poems*. New York: Davidson Press, 1955.

Steinman, David B., and Sara Ruth Watson. *Bridges and Their Builders*. Rev. ed. New York: Dover, 1957. First published 1941 by G. P. Putnam's Sons.

Taylor, George Rogers. *The Transportation Revolution, 1815–1860*. Economic History of the United States, vol. 4. New York: Holt, Rinehart and Winston, 1951.

Troitsky, M. S. *Planning and Design of Bridges*. New York: John Wiley & Sons, 1994.

Turner, George Edgar. *Victory Rode the Rails: The Strategic Place of Railroads in the Civil War*. Lincoln: University of Nebraska Press, 1992. First published 1953.

Waddell, J. A. L. *Bridge Engineering*. 2 vols. New York: John Wiley & Sons, 1916.

Whipple, Squire. *An Elementary and Practical Treatise on Bridge Building*. Rev. ed. New York: D. Van Nostrand, 1872.

Whitney, Charles S. *Bridges: A Study in Their Art, Science and Evolution*. New York: William Edwin Rudge, 1929. Reprinted as *Bridges: Their Art, Science and Evolution*, New York: Greenwich House, 1983.

WORKS ON INDIVIDUAL BRIDGES, ENGINEERS, AND MANUFACTURERS

Arwade, Sanjay, R. Liakos Ariston, and Thomas Lydigsen. "Structural Systems of the Bollman Truss Bridge at Savage, Maryland." *APT Bulletin: Journal of Preservation Technology* 37, no. 1 (2006): 27–35.

Chamberlin, William P. "The Cleft-Ridge Span: America's First Concrete Arch." *IA, The Journal of the Society for Industrial Archeology* 9, no. 1 (1983): 29–44.

Doig, Jameson W., and David P. Billington. "Ammann's First Bridge: A Study in Engineering, Politics, and Entrepreneurial Behavior." *Technology and Culture* 35 (July 1994): 537–70.

Farnham, Jonathan. "Staging the Tragedy of Time: Paul Cret and the Delaware River Bridge." *Journal of the Society of Architectural Historians* 57, no. 3 (September 1998): 258–79.

Fraser, Clayton B. [Biography and works of George S. Morison] in "Nebraska City Bridge." Historic American Engineering Record (HAER), HAER, NEB,66-NEBCI,5-, Library of Congress, 1986.

Hadlow, Robert W. *Elegant Arches, Soaring Spans: C. B. McCullough, Oregon's Master Bridge Builder.* Corvallis: Oregon State University Press, 1999.

Jackson, Robert W. *Rails across the Mississippi: A History of the St. Louis Bridge.* Champaign: University of Illinois Press, 2001.

Kemp, Emory L. "Charles Ellet, Jr. and the Wheeling Suspension Bridge." In *Proceedings of an International Conference on Historic Bridges to Celebrate the 150th Anniversary of the Wheeling Suspension Bridge*, edited by Emory Kemp, 15–32. Morgantown: West Virginia University Press, 1999.

Kranakis, Eda. "Fixing the Blame: Organization Culture and the Quebec Bridge Collapse." *Technology and Culture* 45 (July 2004): 487–518.

Lewis, Gene D. *Charles Ellet, Jr.: The Engineer as Individualist, 1810–1862.* Urbana: University of Illinois Press, 1968.

McCullough, David. *The Great Bridge.* New York: Simon & Schuster, 1972.

Miller, Carol Poh. "The Rocky River Bridge: Triumph in Concrete." *IA, The Journal of the Society for Industrial Archeology* 2, no. 1 (1976): 47-58.

Phoenix Bridge Company. *Album of Designs of The Phoenix Bridge Company, Phoenixville Bridge Works.* Philadelphia: J. B. Lippincott & Co., 1885.

Rastorfer, Darl. *Six Bridges: The Legacy of Othmar H. Ammann.* New Haven, CT: Yale University Press, 2000.

Robb, Frances C. "Cast Aside: The First Cast-Iron Bridge in the United States." *IA, The Journal of the Society for Industrial Archeology* 19, no. 2 (1993): 48–62.

Simmons, David A. "Bridge Building on a National Scale: The King Iron Bridge and Manufacturing Company." *IA, The Journal of the Society for Industrial Archeology* 15, no. 2 (1989): 23–39.

Steinman, David B. *The Builders of the Bridge: The Story of John Roebling and His Son.* 2nd ed. New York: Harcourt, Brace, 1950.

Talese, Gay. *The Bridge.* New York: Harper & Row, 1964.

Trachtenberg, Alan. *Brooklyn Bridge, Fact and Symbol.* 2nd ed. Chicago: University of Chicago Press, 1979.

Vogel, Robert M. *The Engineering Contributions of Wendel Bollman.* United States, National Museum, Bulletin 240; Contributions from the Museum of History and Technology. Washington, DC: Smithsonian Institution, 1964.

_____. *Roebling's Delaware & Hudson Canal Aqueducts.* Smithsonian Studies in History and Technology no. 10. Washington, DC: Smithsonian Institution, 1971.

Zink, Clifford W., and Dorothy White Hartman. *Spanning the Industrial Age: The John A. Roebling's Sons Company, Trenton, New Jersey, 1848–1974.* Trenton, NJ: Trenton Roebling Community Development Corporation, 1992.

TYPOLOGICAL WORKS

Billington, David P., and Aly Nazmy. "History and Aesthetics of Cable-Stayed Bridges." *Journal of Structural Engineering* 117, no. 10 (October 1990): 3103–34.

Boothby, Thomas. "Designing American Lenticular Truss Bridges 1878–1900." *IA, The Journal of the Society for Industrial Archeology* 30, no. 1 (2004): 5–17.

Conwill, Joseph D. *Covered Bridges across North America.* Osceola, WI: Motorbooks International, 2004.

Darnell, Victor. "Lenticular Bridges from East Berlin, Connecticut." *IA, The Journal of the Society for Industrial Archeology* 5, no. 1 (1979): 19–32.

――――. "The National Bridge and Iron Works and the Original Parker Truss." *IA, The Journal of the Society for Industrial Archeology* 24, no. 2 (1998): 5–20.

Emperger, Fritz von. "The Development and Recent Improvement of Concrete-Iron Highway Bridges." *Transactions—American Society of Civil Engineers* 31 (April 1894): 437–88.

Figg, Linda, and W. Denney Pate. "Precast Concrete Segmental Bridges—America's Beautiful and Affordable Icons." *PCI Journal* 49, no. 5 (September–October 2004): 26–38.

Gasparini, Dario A., and David A. Simmons. "American Truss Bridge Connections in the 19th Century, I: 1829–1850, II: 1850–1900." *Journal of Performance of Constructed Facilities* 11, no. 3 (August 1997): 119–29, 130–40.

Gasparini, Dario A., Justin M. Spivey, Stephen G. Buonopane, and Thomas E. Boothby. "Stiffening Suspension Bridges." In *Proceedings of an International Conference on Historic Bridges to Celebrate the 150th Anniversary of the Wheeling Suspension Bridge*, edited by Emory Kemp, 105–16. Morgantown: West Virginia University Press, 1999.

Guise, David. "Elusive American Truss Bridges: Thacher's Truss." *Society for Industrial Archeology Newsletter* 30, no. 2 (2001): 14–15.

Hovey, Otis Ellis. *Movable Bridges.* 2 vols. New York: John Wiley & Sons, 1926–1927.

Luten, Daniel. *Reinforced Concrete Bridges.* Indianapolis: Hollenbeck Press, 1917.

McKee, Brian J. *Historic American Covered Bridges.* New York: American Society of Civil Engineers and Oxford University Press, 1997.

Melan, Josef. *Plain and Reinforced Concrete Arches.* Translated by D. B. Steinman. New York: John Wiley & Sons, 1917.

Scott, Richard. *In the Wake of Tacoma: Suspension Bridges and the Quest for Aerodynamic Stability.* Reston, VA: ASCE Press, 2001.

Simmons, David A. "Bridges and Boilers: Americans Discover the Wrought-Iron Tubular Bowstring Bridge." *IA, The Journal of the Society for Industrial Archeology* 19, no. 2 (1993): 63–79.

――――. "'Light, Aerial Structures of Modern Engineering': Early Suspension Bridges in the Ohio Valley." In *Proceedings of an International Conference on Historic Bridges to Celebrate the 150th Anniversary of the Wheeling Suspension Bridge*, edited by Emory Kemp, 73–86. Morgantown: West Virginia University Press, 1999.

Steinman, David B. *A Practical Treatise on Suspension Bridges: Their Design, Construction and Erection.* New York: John Wiley & Sons, 1922.

CITY, STATE, AND REGIONAL STUDIES

Allen, Richard Sanders. *Covered Bridges of the Midwest.* Brattleboro, VT: Stephen Greene Press, 1970.

_____. *Covered Bridges of the Northeast.* Brattleboro, VT: Stephen Greene Press, 1957.

_____. *Covered Bridges of the South.* Brattleboro, VT: Stephen Greene Press, 1970.

Axline, Jon. *Monument above the Water: Montana's Historic Highway Bridges, 1860–1956.* Helena: Environmental & Hazardous Waste Bureau, Montana Department of Transportation, 1993.

Clouette, Bruce, and Matthew Roth. *Connecticut's Historic Highway Bridges.* Hartford: Connecticut Department of Transportation, 1991.

_____. *Historic Highway Bridges of Rhode Island.* Providence: Rhode Island Department of Transportation, 1990.

Cooper, James L. *Artistry and Ingenuity in Artificial Stone: Indiana's Concrete Bridges, 1900–1942.* Greencastle, IN: J. L. Cooper, 1997.

_____. *Iron Monuments to Distant Posterity: Indiana's Metal Bridges, 1870–1930.* N.P., n.p: 1987.

Deibler, Dan Grove, and Paula A. C. Spero. *A Survey and Photographic Inventory of Metal Truss Bridges in Virginia, 1865–1932.* 9 vols. Charlottesville: Virginia Highway and Transportation Research Council, 1975–1982.

Draper, Joan, consultant; Naomi Draper, editor; Schumacher & Bowman, Inc., consulting engineers. *Chicago Bridges.* Chicago: City of Chicago, Department of Public Works, 1984.

Fraser, Clayton B. *Historic Bridges of Colorado.* Denver: Department of Highways, 1986.

Hess, Jeffrey A., and Robert M. Frame III. *Historic Highway Bridges in Wisconsin.* 3 vols. Madison: Wisconsin Department of Transportation, 1986–1998.

Holstine, Craig, and Richard Hobbs. *Spanning Washington: Historic Highway Bridges of the Evergreen State.* Pullman: Washington State University Press, 2005.

Hyde, Charles K. *Historic Highway Bridges of Michigan.* Detroit: Wayne State University Press, 1993.

Kidney, Walter C. *Pittsburgh's Bridges: Architecture and Engineering.* Pittsburgh: Pittsburgh History & Landmarks Foundation, 1999.

McCullough, Robert. *Crossings: A History of Vermont Bridges.* Barre: Vermont Historical Society and Vermont Agency of Transportation, 2005.

Mikesell, Stephen D. *Historic Highway Bridges of California.* Sacramento: California Department of Transportation, 1990.

Myer, Donald Beekman. *Bridges and the City of Washington.* Washington, DC: U.S. Commission of Fine Arts, 1974.

Pennsylvania Historical and Museum Commission and Pennsylvania Department of Transportation. *Historic Highway Bridges of Pennsylvania.* Harrisburg: Commonwealth of Pennsylvania, 1986.

Quivik, Fredric L. "Montana's Minneapolis Bridge Builders." *IA, The Journal of the Society for Industrial Archeology* 10, no. 1 (1984): 35–54.

Rae, Steven R., Joseph E. King, and Donald R. Abbe. *New Mexico Historic Bridge Survey*. Santa Fe: New Mexico State Highway and Transportation Department, 1987.

Reed, Henry Hope, Robert M. McGee, and Esther Mipaas. *Bridges of Central Park*. New York: Greensward Foundation, 1990.

Reier, Sharon. *The Bridges of New York*. Mineola, NY: Dover, 2000. First published 1977 by Quadrant Press.

Shank, William H. *Historic Bridges of Pennsylvania*. 4th ed. York, PA: American Canal & Transportation Center, 1997.

Smith, Dwight A., James B. Norman, and Pieter T. Dykman. *Historic Highway Bridges of Oregon*. 2nd ed. Portland: Oregon State Historical Society Press, 1989.

Spero, Paula A. C. *A Survey and Photographic Inventory of Concrete and Masonry Arch Bridges in Virginia*. Charlottesville: Virginia Highway & Transportation Research Council, 1984.

White, Joseph, and M. W. von Bernewitz. *The Bridges of Pittsburgh*. Pittsburgh: Cramer Printing & Publishing, 1928.

GLOSSARY

ABUTMENT. The structure upon which the end of a span rests (see 3-134).

ANCHORAGE. The part of a suspension bridge to which the ends of the cables are attached (see 5-004).

AQUEDUCT. A bridge carrying water in pipes or an open channel for irrigating, providing drinking water, or conveying a canal across a river or other obstacle (see 2-022, 5-024).

ARCH. A curved structure whose load is distributed to the points of support radially, primarily by compression. Arches may be fixed so that the structure acts as a monolithic unit (see 2-001) or hinged at each end and sometimes at the crown as well (see 2-067) to allow segments to adjust to changes in temperature and shifting loads.

ASHLAR MASONRY. Stone that is dressed square and laid in regular courses (see 2-013). Ashlar facing refers to such masonry work cladding a structure.

BACKSTAY. A cable or beam in tension supporting a structure (such as a segment of an arch) during construction (see 2-080).

BASCULE BRIDGE. A type of movable bridge consisting of one or more counterweighted leaves (deck sections) that can be tilted upward about a horizontal pivot (see 4-073).

BEAM. A rigid horizontal structure (see 1-002, 1-072).

BEARING. The connection between the end of a span and the abutment. Bearings can be fixed or movable to permit adjustments in the structure due to temperature changes and shifting loads (see 3-173).

BEDSTEAD PONY TRUSS. A pony truss with vertical end posts extending below the deck into the abutments (see 3-210).

BENT. A rigid frame supporting the superstructure of a bridge. Also commonly called a trestle (see 1-013).

BOX GIRDER. A steel or reinforced concrete beam shaped to form a rectangular tube (see 1-074).

CABLE SADDLE. A device on the tower of a suspension bridge that bears the main cables (see 5-018)

CAISSON. A watertight, box-like structure with an open bottom that can be submerged to allow workers access to riverbeds to excavate foundations for piers (see 3-285).

CAMELBACK TRUSS. A form of Parker truss with the top chord composed of five segments (see 3-220).

CANTILEVER. A structural element, such as a beam, extending beyond its point of support (see 3-371).

CAST IRON. A relatively brittle alloy of iron with high carbon content shaped by being poured in molten form into molds. See also *Steel* and *Wrought iron*.

CENTERING. The temporary structure upon which an arch is constructed (see 2-019). The more general term for such structures is "falsework."

CHORD. The upper or lower primary member of a truss (see 3-014).

COMPRESSION. The pushing stress within a structure.

CONCRETE. A paste of sand, water, and cement bonded with aggregate (usually gravel or crushed stone) that hardens into a rock-like material with great compressive strength. See *Reinforced concrete*.

CONSIDÈRE HINGE. A connection in reinforced concrete construction invented by the French engineer Armand Considère (1841–1914) that conveys movement through the flexibility of bunched reinforcing bars rather than a pinned hinge (see 2-173).

CONTINUOUS SPAN. A span with intermediate supports (see 1-004).

CROWN. The top of an arch. Also, the highest point on a roadway shaped to drain water to the sides.

DEAD LOAD. The inherent weight of a structure considered without the action of dynamic live loads.

DECK. The surface of a bridge carrying traffic.

DECK TRUSS SPAN. A bridge span in which the trusses support the deck from below (see 3-216).

DEFLECTION. The bending of a structure under a load.

DEFLECTION THEORY. Theory regarding the means of analyzing the structural behavior of suspension bridges developed by Josef Melan (1853–1941) in 1888 and introduced to the United States by Leon Moisseiff (1872–1943). Widely used by engineers in the first half of the twentieth century.

EYEBAR. A bar of wrought iron or steel with a hole at each end that can be attached to other structural members like links in a chain (see 5-070).

FALSEWORK. The temporary supports upon which an arch, truss, or other structure is constructed. See *Centering*.

FENDER. The structure protecting the piers of a bridge from collisions by waterborne objects (see 4-012).

GIRDER. A large beam serving as a primary load-bearing member in a structure (see 1-023).

I-BEAM. The common name for a beam of rolled wrought iron, steel, or cast reinforced concrete shaped with a cross-section resembling the letter "I" (see 1-003).

KEYSTONE. The voussoir (see below) at the top of an arch that locks the other voussoirs into place (see 2-009).

LIVE LOAD. The dynamic loads acting on a structure, such as (on a bridge) traffic, wind, and the weight of ice.

MEMBER. A component of a structure.

PHOENIX COLUMN. A type of column patented by the Phoenix Iron Works in 1862 and widely used in the late nineteenth century consisting of rolled, flanged channels joined by rivets (see 3-329).

PIER. A vertical structure supporting a bridge between the abutments (see 3-378). In common usage, it is a more massive structure than a column or bent.

PILE. A long column of wood, steel, or concrete set deeply in the ground as part of the foundation of a structure (see 1-010).

PIN CONNECTION. A method of joining the members of a truss by using short metal rods, similar to a door hinge (see 3-102).

PLATE GIRDER. A beam built up from small components rather than rolled to form a given shape and dimension (see 1-030).

PONY TRUSS SPAN. A bridge span with the deck supported between a pair of trusses that are connected only at the bottom chords (see 3-195).

PORTAL. The assembly of columns and upper cross ties at the end of a through truss (see 3-380).

POSTTENSIONING. A method of prestressing a reinforced concrete structure by embedding conduit containing steel wires within the concrete. After the concrete has cured, powerful jacks pull the wires in such a way that compressive stress is introduced into the structure.

PRESTRESSING. The application of compressive stress to a structure to improve its response to live loads (see 2-169). A simple form of prestressing is the introduction of a positive camber, or arching, to a bridge so that under the weight of traffic it deflects to a neutral (level) position. Prestressing may be accomplished by posttensioning or pretensioning.

PRETENSIONING. The application of compressive stress in a reinforced concrete structure by tensioning steel wires before the concrete is poured. Once the concrete has cured, the wires are released and impose a compressive stress on the concrete as they try to retract to a neutral position.

PYLON. The term given to monumental towers found usually in pairs at the approaches to or on the deck of a commemorative bridge (see 2-084).

QUADRANGULAR TRUSS. A truss with intersecting diagonal members that form webs having a quadrangular pattern. The term may be used to describe a double-intersection truss (see 3-274, 3-332).

RAINBOW ARCH. The term given to the distinctive reinforced-concrete-arch bridges designed by James Barney Marsh (1856–1936) and similar bridges with pronounced arches rising above the deck (see 2-153).

REINFORCED CONCRETE. Concrete embedded with steel, which adds tensile strength.

RIGID FRAME. A structure designed so that all its members act as a single unit. Its connections are fixed by techniques such as riveting or welding (see 1-066).

RING ARCH. The outer course of stone forming an arch (see 2-019).

RIVET. A metal fastener with a head on one end like a bolt that is inserted through holes in the members being joined and hammered to form a tightly fitting head on the other end (see 3-351).

RUBBLE MASONRY. Masonry construction using irregularly shaped stones laid randomly or in courses (see 2-009).

SHOE. The device in which the end of a span rests at the bearing (see 3-186).

SIMPLE SPAN. A span supported at each end without intermediary supports (see 1-004).

SKEW. The arrangement of a bridge in which the superstructure is not perpendicular to the substructure (see 3-192).

SKEWBACK. The end of an arch inclined to meet the bearing and abutment (see 2-094).

SPAN. The space between two supports. Also, the structure extending between two supports. With respect to bridges, "span" sometimes refers to the overall length of the structure; the term "clear span" is used to describe the distance between supports.

SPANDREL. On an arch bridge, the triangular area between the arch and the deck (see 2-038).

SPANDREL COLUMNS. Columns supporting the deck of a steel or reinforced concrete arch bridge that bear on the arch (see 2-168).

STEEL. An alloy of iron with a lower carbon content than cast iron, which gives it higher tensile strength. It can be shaped by casting and rolling.

STRINGER. A beam oriented in the direction of the span (see 1-022).

SUBSTRUCTURE. The abutments, foundations, and piers supporting a span.

SUPERSTRUCTURE. The structure forming the span.

SUSPENDERS. The cables or rods supporting the deck on a suspension bridge and some types of arch bridges (see 5-009). Sometimes called hangers.

TENSION. The pulling stress within a structure.

THROUGH TRUSS SPAN. A span in which the trusses rise high above the bridge deck and are connected at the top and bottom chords (see 3-324).

TIE. A member of a structure working in tension (see 3-136).

TIED ARCH. An arch with its ends connected by a tie, like the string of a bow (see 3-024).

TRESTLE. A beam bridge consisting of a series of relatively short spans supported by bents or piles (see 1-014).

TRUNNION. A horizontal pivot on which the leaf of a bascule bridge rotates (see 4-087, 4-089).

TRUSS. A beam or arch composed of relatively small, mutually reinforcing members usually arranged as a series of triangles (see 3-005). See pages 128–130 for definitions of the truss types surveyed in this book.

VIADUCT. A long, multi-span bridge. The term is commonly applied to long bridges crossing features such as valleys, marshy terrain, and waterways (see 1-035).

VIERENDEEL TRUSS. A type of truss composed of rectangular or trapezoidal panels and rigid connections (see 5-086). Named after Belgian engineer Arthur Vierendeel (1852–1940).

VOUSSOIR. A wedge-shaped block used in constructing a masonry arch (see 2-019).

WEB. The members connecting the top and bottom chords of a truss (see 3-315).

WROUGHT IRON. An alloy of iron with a lower carbon content than cast iron or steel. It is malleable with high tensile strength but is less hard than steel.

INDEX

ABOUT THE CD-ROM

The CD-ROM includes direct links to four of the most useful online catalogs and sites, which you may choose to consult in locating and downloading images included on it or related items. Searching directions, help, and search examples (by text or keywords, titles, authors or creators, subject or location, and catalog and reproduction numbers, etc.) are provided online, in addition to information on rights and restrictions, how to order reproductions, and how to consult the materials in person.

1. The Prints & Photographs Online Catalog (PPOC) (http://www.loc.gov/rr/print/catalogabt.html) contains over one million catalog records and digital images representing a rich cross-section of graphic documents held by the Prints & Photographs Division and other units of the Library. It includes a majority of the images on this CD-ROM and many related images, such as those in the HABS and HAER collections cited below. At this writing the catalog provides access through group or item records to about 50 percent of the Division's holdings.

SCOPE OF THE PRINTS AND PHOTOGRAPHS ONLINE CATALOG

Although the catalog is added to on a regular basis, it is not a complete listing of the holdings of the Prints & Photographs Division, and does not include all the items on this CD-ROM. It also overlaps with some other Library of Congress search systems. Some of the records in the PPOC are also found in the LC Online Catalog, mentioned below, but the P&P Online Catalog includes additional records, direct display of digital images, and links to rights, ordering, and background information about the collections represented in the catalog. In many cases, only "thumbnail" images (GIF images) will display to those searching outside the Library of Congress because of potential rights considerations, while onsite searchers have access to larger JPEG and TIFF images as well. There are no digital images for some collections, such as the Look Magazine Photograph Collection. In some collections, only a portion of the images have been digitized so far. For further information about the scope of the Prints & Photographs online catalog and how to use it, consult the Prints & Photographs Online Catalog *HELP* document.

WHAT TO DO WHEN DESIRED IMAGES ARE NOT FOUND IN THE CATALOG
For further information about how to search for Prints & Photographs Division holdings not represented in the online catalog or in the lists of selected images, submit an email using the "Ask a Librarian" link on the Prints & Photographs Reading Room home page or contact: Prints & Photographs Reading Room, Library of Congress, 101 Independence Ave., SE, Washington, D.C. 20540-4730 (telephone: 202-707-6394).

2. The American Memory site (http://memory.loc.gov), a gateway to rich primary source materials relating to the history and culture of the United States. The site offers more than seven million digital items from more than 100 historical collections.

3. The Library of Congress Online Catalog (http://catalog.loc.gov/) contains approximately 13.6 million records representing books, serials, computer files, manuscripts, cartographic materials, music, sound recordings, and visual materials. It is especially useful for finding items identified as being from the Manuscript Division and the Geography and Map Division of the Library of Congress.

4. Built in America: Historic American Buildings Survey/Historic American Engineering Record, 1933–Present (http://memory.loc.gov/ammem/collections/habs_haer) describes and links to the catalog of the Historic American Buildings Survey (HABS) and the Historic American Engineering Record (HAER), among the most heavily represented collections on the CD-ROM.